NETS & DOORS

SHRIMPING IN SOUTHERN WATERS

NETS & DOORS

SHRIMPING IN SOUTHERN WATERS

JACK LEIGH

WYRICK & COMPANY

A Southern Images Book

Published by Wyrick & Company
P.O. Box 89
Charleston, SC 29402

Set in Linotron Sabon with typositor Bembo display
by G & S Typesetters, Inc.
Printed and bound by Meriden-Stinehour Press

The paper in this book meets the guidelines for permanence
and durability of the Committee on Production Guidelines
for Book Longevity of the Council on Library Resources.

Printed in the United States of America

Library of Congress Cataloging-in-Publication Data

Leigh, Jack, 1948–
Nets & doors : shrimping in southern waters / Jack Leigh.
p.   cm.
"A Southern images book"—T.p. verso.
ISBN 0-941711-12-9 : $45.00
1. Shrimp fisheries—Southern States—Pictorial works.   I. Title.
II. Title: Nets and doors.
QL444.M33L44   1989
779'.9639'543—dc20   89-38030
                                                          CIP

Frontispiece: Blessing of the Fleet

ACKNOWLEDGMENTS

IN CREATING *Nets & Doors,* the enjoyment of making the images was complemented by the great satisfaction I had in meeting so many wonderful people. It was through these many generous individuals that this book was made possible.

I wish to thank the following for their very kind financial support and encouragement:

Mr. and Mrs. Leopold Adler, II

Jane and Edwin Feiler, Jr.

Mr. and Mrs. Beverly E. Leigh

The Darien Telephone Company:
Mary Lou Forsyth, Bessie R. Jackson, Reginald V. Jackson

The Sea Island Company: The A. W. Jones Family

The Palmer & Cay/Carswell Company: John E. Cay, III

Within the shrimping community, I am indebted to Lawrence and Gay Jacobs, Hunter and Suzanne Forsyth, Howell Boone, Elaine and Robert Earl Knight, Lloyd Knight, Robert Everson, Wesley Dickey, Rabbit Brigdon, Sam Branson, Charles Deshocka, Roy Clark, Paul Clark, Albert Clark, J. D. McPhail, Henry Hill, John Sawyer, John Henry McIver, The Gale Family, Steve Smith, Albert Walker, Scott and Vicki Richardson, Don Miller, Jerry Bowen, Frank Godley, Billy Godley, Frank Tucker, Leonard Crosby, Fulton Love, Hank Henries, Warren Underwood, Charles Rogers, The Dinh Family, Calvin Lang and Bob McKinley.

Special thanks go to the following for their valuable assistance throughout the project: Susan Shipman and Gordon Rogers, the Georgia Department of Natural Resources, Coastal Resources Division; Paul Christian, University of Georgia Marine Extension Service; Reita Rivers and Mac Rawson, University of Georgia Sea Grant Program; Linda O. King, Director, and T. Reed Ferguson, Curator, Museum of Coastal History; Susan Hardwick, Florida Times Union; Woody Woodsides, Director, Brunswick/Golden Isles Chamber of Commerce; Jane Feiler, President, CELSA Learning Systems; Julian Weitz, Vice-president, Friedman's Art Store; William Haynes; Jeannine Cook; Buddy Sullivan, Editor, Darien News; Betty McDaniel, Louise Jones, and Kathi Dailey, Production Coordinators and Designers, University of Georgia Press.

I especially thank Sandra Hudson, Assistant Director and Production and Design Manager, University of Georgia Press, for her care, concern and friendship in bringing my books to fruition; Betty Herrington, who is the best assistant one could hope for, and a trusted and loyal friend; Gunnar Lindquist, financial adviser and friend, whose support and guidance have been invaluable; Karen and Liza Penick, whose support and love were so important during the shooting of the book; Charles L. Wyrick, Jr., President, Wyrick & Company, and my partner in Southern Images Publishing. The fact that Pete Wyrick and I are publishing the first book of our new partnership is a dream come true. From the beginning of our relationship, when we published *Oystering: A Way of Life*, it has been our mutual desire to depict the South in such a way that its authentic beauty is revealed. I feel grateful for the opportunity to continue our work together; Robert McAlister, my close friend and confidant, whose insights into my life and work have been invaluable; Haywood Nichols, fellow artist, whose sculpture inspires me and whose friendship I treasure; to all my students with whom I grow in a deeper understanding and appreciation of photography; and Susan Miller, President and Director, Southern Images Photographic Workshops. It is my great fortune to have Susan as a partner in our teaching of photography, and I am especially blessed by her sharing with me her love and vision.

To my mother, Mary Pindar Leigh, who has given me through her art and her painting, the joy of life and the magic of creating images; and to my father, Beverly Eugene Leigh, who has given me through his strength and his encouragement, the will to do my work and the faith to follow my heart.

SHRIMPING IN SOUTHERN WATERS

ONE OF MY FONDEST MEMORIES OF CHILDHOOD was the car ride I would take with my parents after Sunday school. My favorite trip was to a nearby fishing village called Thunderbolt. We would park under the sprawling oaks on a high bluff overlooking the river. From this vantage point we could watch the shrimp boats, draped with their sea-green nets, come and go. It was an enchanting sight. The bluff was always crowded on Sundays with those like us who had come to watch and to breathe in the deep aroma of the salt marsh, and others who had set up their easels and tripods to paint and to photograph. Years later I would tag along with my mother, who had become an artist, and watch as she positioned her easel on the bluff and painted the rugged, sea-battered boats and the rickety old docks.

Those early images of shrimp boats have been with me ever since, and as a photographer I have always wanted to return to Thunderbolt to make my own pictures of those boats and the world of shrimp fishing that have been such an integral part of our Southern heritage.

As the years passed, the view from the high bluff overlooking the river revealed fewer traditional shrimp boats and old docks, and more sleek pleasure boats and new marinas. Thunderbolt, where shrimping had originated on the Georgia coast in the early part of this century, was undergoing dramatic changes. The docks were being sold and the shrimp boats were disappearing. Families, who for generations depended on the sea for their livelihood, were having to find other places to live and work, or were giving up shrimping completely.

Today, there is only a vestige of the Thunderbolt I knew as a child. If I wanted to take my pictures of the shrimp boats, the shrimp fishermen, and the old docks, I knew that I must do it now.

Over the course of two years, I immersed myself in the world of these shrimp fishermen. Theirs is a life characterized by hard work and fierce individualism. They work long, arduous hours, and they are proud and independent people.

I began each day well before dawn, meeting the captain at the shrimp docks. The strikers were already on board, carefully pulling through the nets, searching for any tears and mending them with net needles and twine. Chains, ropes and winches were checked and prepared for the day's work.

The quietness of the surrounding marsh was abruptly shattered as the shrimp boat's engines were started. We made our way through dark waterways to the open sea; diesel fumes and the pungent smells of salt and strong coffee filled the air.

As the captain neared his chosen fishing grounds, the outriggers were let down, spreading out from the boat like giant metal wings. A rusting and creaking winch was brought to life as the water-logged, wooden trawl doors were set out. The bag at the end of the net was tied shut to capture the elusive shrimp.

The sea around us rolled black and eerie as the nets were thrown over the sides. Work lights from nearby shrimp boats provided a welcomed camaraderie during that lonely and murky time before sunrise.

At dawn, the nets were lowered to the bottom, the drag nearly bringing the boat to a halt. A thick whiff of frying bacon cut through the early morning chill as the strikers cooked up a hearty breakfast. The captain flipped on the radio and called across to other boats pulling their nets in the dim light. The conversations were always the same—Where are the shrimp?

By instinct and experience, the captain navigated his boat, usually fishing the same waters every day. He was familiar with the bottoms, knowing that there were no ancient anchors or skeletons of sunken ships there to snag his nets. The captain also kept his "snag book," containing years of jottings indicating treacherous bottoms to avoid.

During the course of a drag, a small "try-net" was set out between the large nets. The strikers pulled this net about every twenty minutes. The captain would put the boat on automatic pilot by tying a rope to the wheel, and come back on deck to count the number of shrimp caught in this net. A good count meant we were in shrimp and we would hold our course; a skimpy catch told the captain to head for different waters.

Anticipation swelled as the huge nets were finally pulled from the sea and the contents spilled on deck. The sight was startling, as a profusion of shrimp and unknown sea creatures slithered, wiggled, and flopped about. The strikers quickly pulled up small wooden stools to this writhing mound and with seasoned hands began culling and heading the shrimp, returning everything else to the sea. Sea gulls swarmed, screeching and feasting. The shrimp were sorted by size and iced down in large storage bins beneath the deck.

Throughout the day the nets were dropped and hauled in several times. As evening approached, the nets were hauled in one last time, the captain and crew hoping for that one big catch.

Trailed by the ever present flock of sea gulls, we "headed for the hill." The outriggers were folded up once more, showing stark against the pale sky of nightfall. Sea-heavy nets were hoisted above the deck, shaken free of any accumulated debris, and hung to dry. The deck was hosed down and the boat was made ready for the next day when the sea and the shrimp would beckon once again.

NETS & DOORS

SHRIMPING IN SOUTHERN WATERS

Navigating

Brothers: Captain Frank Godley, Striker Billy Godley

Setting out the nets

Captain Wesley Dickey working the winch

Untangling the gear

Captain John McPhail chatting on the radio

Nets & Doors

Hauling in the nets

Sea gulls

Tripping the bag

Captain Robert Everson

Striker preparing to cull shrimp

Strikers culling shrimp

The catch

Between hauls

Striker Frank Tucker taking a coffee break

Captain John Sawyer hauling back the gear

Decking the doors

Nets

Captain Henry Hill

Striker Alfred Walker

Captain Steve Smith

Portside

Anchored-up

Captain Jerry Bowen, Striker Don Miller

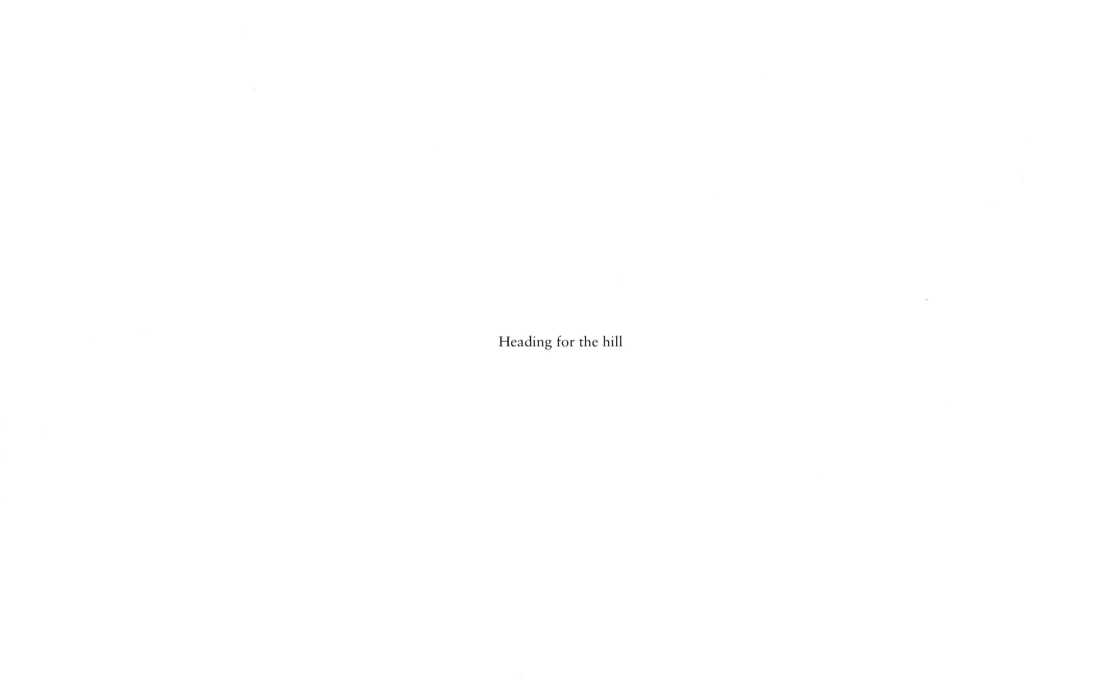

Heading for the hill

Staples

Tying up

Shrimp boat dock

Captain Hunter Forsyth, Dock Manager Suzanne Forsyth

Sunken shrimp boat

Weathered dock

Striker couple: Scott and Vicki Richardson

Shrimping family: Captain Darwin Gale, Captain Dave Lester Gale, Sr.,
Captain Dave Lester Gale, Jr., Darrell, Darwin, Jr. and Davey Gale

The railway

Lloyd Knight working on the railway

Sam Branson mending the nets

Captain John Henry McIver repairing his nets

Shrimp boat mold

Waiting for the shrimp boats to return

Day after day, I went to sea with the shrimp fishermen. Some days were good; most days were bad. Yet, the spirit of the fishermen always looked forward to tomorrow and the hope of a bountiful catch.

ABOUT THE AUTHOR

THE SOCIALLY CONCERNED PHOTOGRAPHER has played a prominent role in the history of American photography since the late 19th century. The pioneering documentary photographers of the American West suggested the impending plight of native Americans and the fragility of our wilderness areas. Lewis Hine and the photographers of the WPA years left indelible records of their respective eras and changed the way that we thought about our country. The fading history and passing traditions of the South are being chronicled today by Jack Leigh, a documentary photographer based in Savannah, Georgia.

Returning home after extended travel enabled Leigh to focus his efforts on the land, people, and traditions that he knew and respected. His first extended project concentrated on the lives of oystermen along the barrier islands of the South Carolina coast. After two years' work on the oyster boats and in the shucking houses and steam factories, he assembled his first book, *Oystering: A Way of Life,* which was published by the Carolina Art Association.

Next, Leigh devoted another two years to a photographic journey on Georgia's Ogeechee River, tracing the river from its source in a field in northeast Georgia to its confluence with the Atlantic Ocean. While *Oystering* was primarily a book of photographs, with a foreword by James Dickey and an essay by Leigh, *The Ogeechee: A River and Its People* included extended descriptions and narratives interspersed with the photographs. *The Ogeechee* was published by The University of Georgia Press as part of the Wormsloe Foundation Publications series.

Leigh has been following the lives of the shrimp fishermen on Georgia's coastal waters since 1986. As technology replaces manual labor and resort developments replace traditional oystering and shrimping, the lives and fortunes of the watermen and river dwellers are being changed forever. Just five years after the publication of *Oystering,* the barges and bateaux are beached, the shucking house and cannery are closed. Leigh's photographs are all that remain of a vital element in the cultural history of the South.

Jack Leigh continues to pursue the passing traditions and pastimes of the South: one of the last black baseball leagues, drive in movie theaters, and the South's ever-changing rural communities, peanut farms, and port cities. While his approach is documentary, his keen sense of composition and his technical expertise in the darkroom will establish Leigh as an artist who will one day be compared to Walker Evans, Robert Frank, Dorothea Lange and other venerated American photographers.

Thailand

Heidi Fröhlich
Katja Sassmanshausen
Thilo Scheu

Bangkok

Koh Kood

Koh Phi Phi Leh

Railay Beach, Krabi

Sam Phan Bok, Ubon Ratchathani

Muang Tham, Buriram

Índice · Contents · Sommaire · Inhalt · Indice · Inhoud

20 Introduction
Einleitung
Introducción
Introduzione
Inleiding

24 Chiang Mai

52 Chiang Rai

66 Traditions
Brauchtum
Costumbres
Tradizioni
Gebruiken

**68 Mae Hong Son,
Tak & Lamphun**

84 Phayao, Nan & Phrae

98 Lampang

110 Spirit houses
Maisons des esprits
Geisterhäuschen
Casas fantasmas
Case degli spiriti
Geestenhuizen

112 Sukhothai

**126 Kamphaeng Phet,
Phitsanulok &
Phetchabun**

146 Loei & Nong Khai

160 Elephants
Éléphants
Elefanten
Elefantes
Elefanti
Olifanten

**162 Udon Thani, Sakon
Nakhon & Nakhon
Phanom**

172 Ubon Ratchathani

**184 Nakhon Ratchasima,
Buriram &
Nakhon Nayok**

194 Monks
Moines
Mönche
Frailes
Monaci
Monniken

196 Lopburi & Ayutthaya

218 Kanchanaburi

250 Bangkok

284 Culinary
Gastronomie
Kulinarisches
Gastronomía
Gastronomia
Culinair

**286 Samut Sakhon,
Samut Prakan,
Nakhon Pathom &
Samut Songkhram**

296 Chonburi

320 Rayong

332 Chantaburi & Trat

**350 Ratchaburi &
Phetchaburi**

370 Temples
Tempel
Templos
Templi
Tempels

**372 Prachuap Khiri Khan,
Chumphon & Ranong**

**394 Surat Thani & Ang
Thong Marine
National Park**

**428 Phang Nga &
Similan Islands**

450 Phuket

466 Buddhas
Bouddhas
Budas
Boeddha's

**468 Krabi, Koh Lanta
& Phi Phi Islands**

**492 Nakhon Si Thammarat
& Songkhla**

504 Trang & Satun

518 Map
Carte
Landkarte
Mapa
Carta
Landkaart

520 Index

523 Photo Credits

Thailand

Buddhist monasteries and golden temples, gorgeous sandy beaches and turquoise waters, bizarre rock formations and pristine islands, mist covered jungle villages and lively cities, unique world heritage sites and a varied cuisine to fall in love with— Southeast Asian Thailand enchants with its unique diversity, a country that arouses longings and fulfils dreams.

The kingdom, which is often called the "land of smiles" because of the friendliness of its people, is home to around 69 million inhabitants and borders on Myanmar, Laos, Cambodia and Malaysia. The north, with its provinces of Chiang Mai and Chiang Rai, is dominated by mountains up to 2600 m high, with dense forests and valleys with flowing rivers. Northern Thailand is the home of ethnic minorities and cultural diversity, and is an Eldorado for hikers and nature lovers. The country has dozens of national parks in which you can see wild elephants, howler monkeys hanging from the trees, or with a little luck, a tiger from afar. As an exciting contrast to peace and solitude, a detour to Thailand's capital Bangkok is recommended. Chaotic traffic and bustle meet royal palaces and peaceful monks: An incomprehensible metropolis full of contrasts. Further south, palm-fringed beaches and fairytale islands bring a smile to every visitor's face, and the magnificent underwater world makes the hearts of divers beat faster.

Thaïlande

Des monastères bouddhistes et des temples dorés, de magnifiques plages de sable fin et des eaux turquoises, d'étranges formations rocheuses et des îles vierges, des villages de jungle couverts de brume et des villes animées, des sites uniques du patrimoine mondial et une cuisine si variée qu'on peut en tomber rapidement amoureux – en Asie du Sud-Est, la Thaïlande enchante par la diversité unique de ses attraits.

Le royaume, que l'on appelle souvent la « terre des sourires » en raison de la gentillesse des Thaïlandais, compte environ 69 millions d'habitants et partage ses frontières avec le Myanmar, le Laos, le Cambodge et la Malaisie. Le nord, avec les provinces de Chiang Mai et Chiang Rai, est dominé par des paysages montagneux atteignant les 2 600 m d'altitude, des forêts denses et des vallées traversées de rivières. La région, habitat de nombreuses minorités ethniques, est un foyer de diversité culturelle mais aussi un eldorado pour les randonneurs et les amoureux de la nature. Le pays compte des dizaines de parcs nationaux où l'on peut admirer des éléphants sauvages, des singes hurleurs suspendus dans les arbres voire, avec un peu de chance, des tigres. Pour contraster avec cette paix et cette nature sauvage, il est recommandé de faire un détour par Bangkok, la capitale de la Thaïlande. La circulation chaotique et l'agitation y cohabitent avec les palais royaux et les moines paisibles. Une métropole pleine de contrastes, parfois incompréhensible. Plus au sud, les plages bordées de palmiers et les îles féeriques donnent le sourire à chaque visiteur et le magnifique monde sous-marin fait battre plus vite le cœur des plongeurs.

Thailand

Buddhistische Klöster und goldene Tempel, traumhafte Sandstrände und türkisfarbenes Wasser, bizarre Felsformationen und makellose Inseln, nebelverhangene Dschungeldörfer und quirlige Städte, einzigartige Welterbestätten und eine abwechslungsreiche Küche zum Verlieben – das südostasiatische Thailand verzaubert mit seiner einmaligen Facettenvielfalt. Ein Land, das Sehnsüchte weckt und Träume erfüllt.

Das gerne aufgrund der Freundlichkeit seiner Menschen als „Land des Lächelns" titulierte Königreich beherbergt rund 69 Millionen Einwohner und grenzt an die Länder Myanmar, Laos, Kambodscha und Malaysia. Den Norden mit den Provinzen Chiang Mai und Chiang Rai dominieren bis zu knapp 2600 m hohe Berge, dichte Wälder und von Flüssen durchzogene Täler. Nordthailand ist der Lebensraum ethnischer Minderheiten, Heimstätte kultureller Vielfalt und ein Eldorado für Wanderer und Naturliebhaber. Das Land besitzt dutzende Nationalparks in denen man wilde Elefanten, durch die Bäume hangelnde Brüllaffen oder mit etwas Glück einen Tiger aus der Ferne erspähen kann. Als spannendes Kontrastprogramm zu Ruhe und Einsamkeit empfiehlt sich ein Abstecher in Thailands Hauptstadt Bangkok. Chaotischer Straßenverkehr und geschäftiges Treiben trifft auf königliche Paläste und friedvolle Mönche. Eine unfassliche Millionenmetropole voller Gegensätze. Weiter im Süden zaubern palmengesäumte Strände und feenhaft schöne Inseln jedem Besucher ein Lächeln ins Gesicht und die prächtige Unterwasserwelt lässt das Herz der Taucher höher schlagen.

Krabi

Haew Narok Waterfall, Khao Yai National Park

Tailandia

Monasterios budistas y templos dorados, hermosas playas de arena y aguas turquesas, extrañas formaciones rocosas e islas inmaculadas, pueblos selváticos envueltos en niebla y ciudades animadas, lugares únicos del patrimonio mundial y una variada gastronomía de la que enamorarse: el sudeste asiático de Tailandia maravilla por su diversidad única. Un país que despierta nostalgias y cumple sueños.

El reino, a menudo denominado la "tierra de las sonrisas" por la amabilidad de su gente, tiene alrededor de 69 millones de habitantes y limita con Myanmar, Laos, Camboya y Malasia. El norte, con las provincias Chiang Mai y Chiang Rai, está dominado por montañas de hasta 2600 m de altura, densos bosques y valles atravesados por ríos. El norte de Tailandia es el hábitat de las minorías étnicas, hogar de la diversidad cultural y un Eldorado para excursionistas y amantes de la naturaleza. El país tiene docenas de parques nacionales en los que se pueden ver elefantes salvajes, monos aulladores colgando entre los árboles o, con un poco de suerte, un tigre desde la lejanía. Como contraste de emociones entre la paz y la soledad, se recomienda desviarse a Bangkok, la capital de Tailandia. El tráfico caótico y el ajetreo se encuentran con palacios reales y monjes pacíficos. Una metrópolis incomprensible, llena de contrastes. Más al sur, las playas rodeadas de palmeras y las islas de ensueño dibujan una sonrisa en los rostrosde todos los visitantes y el magnífico mundo submarino hace que el corazón de los buceadores lata más rápido.

Tailandia

Monasteri buddisti e templi dorati, splendide spiagge sabbiose e acque turchesi, bizzarre formazioni rocciose e isole incontaminate, villaggi avvolti nella nebbia della giungla e città frenetiche, luoghi straordinari, riconosciuti come patrimonio dell'umanità, e una cucina variegata da cui farsi prendere per la gola – la Thailandia, nel sud-est asiatico, incanta il visitatore per le sue tante sfaccettature che la rendono unica. Un paese che risveglia i desideri e trasforma i sogni in realtà.

Nel paese al confine con Myanmar, Laos, Cambogia e Malaysia, una monarchia spesso chiamata la "terra dei sorrisi" per la cordialità della gente, vivono circa 69 milioni di abitanti. Il nord, con le province di Chiang Mai e Chiang Rai, è caratterizzato da montagne alte fino a 2.600 m, fitti boschi e valli attraversate da fiumi ed è un eldorado per gli escursionisti e gli amanti della natura. Queste regioni sono caratterizzate dalla diversità culturale perché qui vivono varie minoranze etniche. Il paese possiede decine di parchi nazionali in cui si possono ammirare elefanti selvatici, scimmie urlatrici appese tra gli alberi o, con un po' di fortuna, qualche tigre in lontananza. Per un programma emozionante, in contrasto con la pace e la solitudine di questi posti, si consiglia una deviazione per Bangkok, la capitale della Thailandia. Qui il traffico caotico e un viavai indaffarato fanno da cornice a palazzi reali e monaci pacifici. Una metropoli difficile da riassumere a parole e piena di contrasti. Più a sud, spiagge bordate di palme e isole piene di magia regalano il sorriso a chiunque le visiti, mentre il magnifico mondo sottomarino fa battere veloce il cuore agli appassionati di immersioni.

Thailand

Boeddhistische kloosters en gouden tempels, prachtige zandstranden en turquoise wateren, bizarre rotsformaties en ongerepte eilanden, nevelige jungledorpen en levendige steden, unieke werelderfgoedlocaties en een gevarieerde keuken om verliefd op te worden – het Zuidoost-Aziatische Thailand betovert met zijn unieke gevarieerdheid. Het is een land dat verlangens oproept en dromen vervult.

Het koninkrijk dat vanwege de vriendelijkheid van zijn bevolking vaak het "land van de glimlach" wordt genoemd, telt ongeveer 69 miljoen inwoners en grenst aan de landen Myanmar, Laos, Cambodja en Maleisië. Het noorden met de provincies Chiang Mai en Chiang Rai wordt gedomineerd door bergen tot 2600 meter hoog, dichte bossen en dalen met rivieren. Noord-Thailand is het leefgebied van etnische minderheden, de thuisbasis van culturele diversiteit en een eldorado voor wandelaars en natuurliefhebbers. Het land heeft tientallen nationale parken waar je wilde olifanten, brulapen die van tak naar tak slingeren en, met een beetje geluk, een tijger kunt bespeuren. Als spannend contrast met al die rust en verlatenheid is een omweg naar de Thaise hoofdstad Bangkok aan te raden. Chaotisch verkeer en grote bedrijvigheid gaan hand in hand met koninklijke paleizen en vredelievende monniken. Een onvoorstelbare miljoenenstad vol tegenstellingen. Verder naar het zuiden brengen met palmbomen omzoomde stranden en sprookjesachtige eilanden een glimlach op het gezicht van alle bezoekers en doet de prachtige onderwaterwereld het hart van duikers sneller kloppen.

Chiang Mai

Elephant

Chiang Mai

The northern province of Chiang Mai presents impenetrable, lush green jungles and remote, slowly flowing rivers. The mountains rise majestically on the horizon, including the highest peak in the country, 2,565 m Doi Inthanon. Richly decorated temples and urban Asian liveliness are the characteristics of the eponymous capital. A region blessed with charming nature and culture.

Chiang Mai

Dans la province septentrionale de Chiang Mai, les jungles verdoyantes et impénétrables sont parcourues par des rivières aux courants lents. Les montagnes s'y dressent majestueusement vers l'horizon, y compris le sommet le plus élevé du pays, le Doi Inthanon, qui culmine à 2 565 m de haut. Dans la capitale du même nom, les temples richement décorés côtoient l'animation urbaine. Une région dotée d'une nature et d'une culture charmantes.

Chiang Mai

Mit undurchdringlichem, sattgrünem Dschungel und abgelegenen, langsam dahinfließenden Flüssen präsentiert sich die nördliche Provinz Chiang Mai. Majestätisch erheben sich die Berge bis zum Horizont, darunter auch der höchste des Landes, der 2565 m hohe Doi Inthanon. Reich verzierte Tempel und urbane asiatische Lebendigkeit sind die Merkmale der gleichnamigen Hauptstadt. Eine Region gesegnet mit reizvoller Natur und Kultur.

Chiang Mai

Chiang Mai

La provincia norteña de Chiang Mai se presenta con impenetrables y exuberantes selvas verdes y ríos remotos que fluyen con calma. Las montañas se elevan majestuosamente hacia el horizonte, incluyendo la más alta del país, el Doi Inthanon, de 2.565 m de altura. Los templos muy ornamentados y la animación urbana asiática son las características de la capital del mismo nombre. Una región bendecida con una naturaleza y una cultura seductoras.

Chiang Mai

La provincia settentrionale di Chiang Mai offre uno spettacolo di giungle verdi e impenetrabili e remoti fiumi che scorrono lenti. Le montagne si ergono maestose all'orizzonte, tra cui la più alta del paese, la Doi Inthanon con i suoi 2.565 m. Templi riccamente decorati e frenesia urbana tipicamente asiatica sono le caratteristiche dell'omonimo capoluogo di questa regione ricca di natura e cultura.

Chiang Mai

De noordelijke provincie Chiang Mai presenteert zich met ondoordringbare, weelderig groene jungles en afgelegen, langzaam stromende rivieren. De bergen rijzen majestueus op tot aan de horizon, waaronder de hoogste berg van het land, de 2565 meter hoge Doi Inthanon. Rijkversierde tempels en stedelijke Aziatische levendigheid kenmerken de gelijknamige hoofdstad. Een regio gezegend met een bekoorlijke natuur en cultuur.

Ho Kham Luang, Royal Pavilion

Rice paddy fields

The mountain world of Chiang Mai

The hills of Chiang Mai province, which extend as far as the border with Myanmar, are an ideal and unique destination for nature lovers. Numerous national parks such as Doi Suthep-Pui at an altitude of over 1600 m are spread throughout the region. On their slopes, the hill tribes living in the north have created artistically shaped rice fields and tea plantations. Atmospheric scenery that changes with the light of day.

Montagnes de Chiang Mai

Les collines de la province de Chiang Mai, qui s'étendent jusqu'à la frontière du Myanmar, sont une destination d'excursion idéale et unique pour les amoureux de la nature. De nombreux parcs nationaux, comme le Doi Suthep-Pui, à plus de 1600 m d'altitude, sont répartis dans toute la région. Sur leurs pentes, les tribus montagnardes vivant dans le nord ont aménagé des rizières et des plantations de thé aux formes artistiques. Un paysage irréel, qui varie selon la lumière.

Bergwelt von Chiang Mai

Die bis an die Grenze von Myanmar reichenden Hügel der Provinz Chiang Mai sind ein idealer und einzigartiger Ausflugsort für Naturliebhaber. Zahlreiche Nationalparks wie der auf einer Höhe über 1600 m liegende Doi Suthep-Pui verteilen sich in der Region. An ihren Hängen haben die im Norden lebenden Bergvölker kunstvoll geformte Reisfelder und Teeplantagen angelegt. Eine sich im Licht des Tages verändernde stimmungsvolle Szenerie.

Tea plantations

El mundo montañoso de Chiang Mai

Las colinas de la provincia de Chiang Mai, que se extienden hasta la frontera de Myanmar, son un destino ideal y único para los amantes de la naturaleza. Numerosos parques nacionales como Doi Suthep-Pui, situado a más de 1600 m, se extienden por toda la región. En sus laderas, las tribus de montaña que viven en el norte han creado campos de arroz y plantaciones de té con formas artísticas. Un paisaje atmosférico que cambia a la luz del día.

Le montagne di Chiang Mai

Le montagne della provincia di Chiang Mai, che si estendono fino al confine con il Myanmar, sono una meta escursionistica straordinaria, ideale per gli amanti della natura. Numerosi parchi nazionali, come quello di Doi Suthep-Pui a oltre 1.600 m di altezza, sono dislocati in tutta la regione. Sulle loro pendici, le popolazioni montane che qui vivono coltivano riso e tè i cui campi e piantagioni disposti in maniera artistica danno vita a uno scenario che cambia durante le ore del giorno al variare della luce.

Bergwereld van Chiang Mai

De heuvels van de provincie Chiang Mai, die zich uitstrekken tot aan de grens van Myanmar, zijn een ideale en unieke bestemming voor natuurliefhebbers. Talrijke nationale parken, zoals het op 1600 meter hoogte gelegen Doi Suthep-Pui, liggen verspreid over de regio. Op de hellingen van dat gebied hebben in het noorden levende bergvolkeren kunstig gevormde rijstvelden en theeplantages aangelegd. Een sfeervol tafereel dat verandert met het licht van de dag.

Yee Peng Lantern Festival

At the Yee Peng Lantern Festival in Chiang Mai, every year thousands upon thousands of lanterns climb into the night sky. The festival is celebrated at the full moon in the twelfth month of the traditional Thai lunar calendar; the date is usually in November. The event is closely connected with the Loy Krathong Festival of Lights, which is celebrated throughout the country.

Festival des lanternes de Yee Pang

Au Yee Peng Festival, à Chiang Mai, des milliers et des milliers de lanternes s'élèvent chaque année dans le ciel nocturne. La fête à lieu à la pleine lune, au douzième mois du calendrier lunaire traditionnel thaïlandais, habituellement en novembre. L'événement est étroitement lié au Festival des lanternes Loy Krathong, célébré dans tout le pays.

Yee Peng Laternen-Festival

Beim Yee Peng Laternen-Festival in Chiang Mai steigen jedes Jahr abertausende Laternen in den Nachthimmel. Es wird bei Vollmond im zwölften Monat des traditionellen thailändischen Mondkalenders begangen. Der Termin liegt meist im November. Die Veranstaltung hängt eng mit dem im ganzen Land zelebrierten Loy Krathong-Lichterfest zusammenhängt.

Festival de linternas de Yee Peng

En el festival de linternas de Yee Peng, en Chiang Mai, miles y miles de linternas suben al cielo nocturno cada año. Se celebra en luna llena en el duodécimo mes del tradicional calendario lunar tailandés. La fecha suele ser noviembre. El evento está estrechamente relacionado con el festival de las luces de Loy Krathong, que se celebra en todo el país.

Festival delle lanterne di Yee Peng

Al festival delle lanterne di Yee Peng, in Chiang Mai, migliaia e migliaia di lanterne si innalzano ogni anno nel cielo notturno. Le celebrazioni hanno luogo nelle notti di luna piena, nel dodicesimo mese del tradizionale calendario lunare thailandese. La data cade di solito a novembre. L'evento è collegato al festival delle luci di Loy Krathong, che si celebra in tutto il paese.

Lampionnenfeest Yee Peng

Tijdens het lampionnenfeest Yee Peng in Chiang Mai stijgen elk jaar duizenden lampionnen op in de nachtelijke hemel. Het wordt in de twaalfde maand van de traditionele Thaise maankalender bij vollemaan gevierd. De datum ligt meestal in november. Het evenement hangt nauw samen met het Krathong-lichtfeest, dat in het hele land gevierd wordt.

Doi Inthanon National Park

Doi Inthanon National Park

Founded in 1954, Doi Inthanon National Park covers an area of almost 500 km². Well-developed hiking and walking trails lead through the unspoilt mountainous landscape with its unique flora. Highlights include several waterfalls, such as the impressive 260 metre high Mae Ya waterfall with its many cascades.

Parc national de Doi Inthanon

Le parc national de Doi Inthanon, fondé en 1954, couvre une superficie de près de 500 km². Des sentiers de balades et de randonnées, bien aménagés, serpentent à travers le paysage montagneux intact et sa flore unique. On peut y admirer plusieurs chutes d'eau, comme l'impressionnante Mae Ya, qui s'écoule de 260 m de haut en de nombreuses cascades.

Nationalpark Doi Inthanon

Der 1954 gegründete Nationalpark Doi Inthanon umfasst eine Fläche von knapp 500 km². Gut ausgebaute Wander- und Spazierwege führen durch die bergige urwüchsige Landschaft mit einer einzigartigen Pflanzenwelt. Highlights sind mehrere Wasserfälle wie der 260 m hohe und imposante Mae Ya-Wasserfall mit seinen vielen Kaskaden.

Mae Ya Falls, Doi Inthanon National Park

Parque nacional Doi Inthanon

El Parque nacional Doi Inthanon, fundado en 1954, tiene una superficie de casi 500 km². Rutas de senderismo bien desarrolladas atraviesan el paisaje montañoso y virgen con su flora única. Destacan varias cascadas como la impresionante cascada de Mae Ya, de 260 m de altura y sus numerosas subvertientes.

Parco nazionale Doi Inthanon

Il Parco nazionale di Doi Inthanon, fondato nel 1954, si estende su una superficie di quasi 500 km². Sentieri escursionistici e percorsi trekking conducono attraverso il paesaggio montano incontaminato, caratterizzato da una flora particolare. Tra i punti salienti vi sono diverse cascate, come quella imponente di Mae Ya, che con un salto di 260 m dà vita a varie cascate più piccole.

Nationaal park Doi Inthanon

Het in 1954 opgericht nationale park Doi Inthanon beslaat een oppervlakte van bijna 500 km². Een uitgebouwd net van wandelroutes leidt door het bergachtige, ongerepte landschap met een unieke flora. Hoogtepunten zijn diverse watervallen zoals de 260 meter hoge en indrukwekkende Mae Ya-waterval met zijn grote hoeveelheid cascades.

Rice terraces, Bong Piang Forest

Wat Phra Singh, Chiang Mai City

Temples in Chiang Mai

The largest and most important temple in Chiang Mai is Wat Phra Singh, founded in 1345. The golden Buddha statue of Phra Sihing, which according to legend comes from Sri Lanka and is said to be almost 3000 years old, is worshipped there. An earthquake destroyed part of Wat Chedi Luang in 1545. The temple was once the home of the Emerald Buddha, highly esteemed in Thailand.

Temples de Chiang Mai

Le temple le plus grand et le plus important de Chiang Mai est le Wat Phra Singh, fondé en 1345, où l'on vénère la statue dorée du Bouddha de Phra Sihing qui, selon la légende, viendrait du Sri Lanka et aurait presque 3000 ans. Dans le centre de Chiang Mai se trouve également le Wat Chedi Luang, en partie détruit par un tremblement de terre en 1545. Ce temple était autrefois la maison du très estimé Bouddha d'émeraude.

Tempel in Chiang Mai

Der größte und bedeutendste Tempel in Chiang Mai ist der 1345 begründete Wat Phra Singh. Verehrt wird dort die goldene Buddha-Statue des Phra Sihing, die der Legende nach aus Sri Lanka stammen und fast 3000 Jahre alt sein soll. Ein Erdbeben zerstörte 1545 einen Teil des Wat Chedi Luang. Einst war der Tempel die Heimstätte des in Thailand hoch geschätzten Smaragd-Buddhas.

Wat Chedi Luang

Templos en Chiang Mai

El templo más grande e importante de Chiang Mai es el Wat Phra Singh, fundado en 1345. La venerada estatua dorada de Buda de Phra Sihing, viene de Sri Lanka y según la leyenda,tiene casi 3000 años. Un terremoto destruyó una parte del temploen 1545. En el pasado, el templo fue el hogar del apreciado Buda de Esmeralda en Tailandia.

Templi di Chiang Mai

Il tempio più grande e più importante di Chiang Mai è il Wat Phra Singh, fondato nel 1345. Qui viene venerata la statua dorata del Buddha di Phra Singh, che secondo la leggenda proviene dallo Sri Lanka e ha quasi 3.000 anni. Un terremoto distrusse una parte del Wat Chedi Luang nel 1545, un tempo dimora del Buddha di Smeraldo, in Thailandia molto venerato.

Tempels in Chiang Mai

De grootste en belangrijkste tempel van Chiang Mai is de in 1345 opgerichte Wat Phra Singh. Het gouden Boeddhabeeld van Phra Sihing, dat volgens de legende uit Sri Lanka komt en naar verluidt bijna 3000 jaar oud is, wordt daar vereerd. In 1545 verwoestte een aardbeving een deel van de tempel Wat Chedi Luang, die ooit de in Thailand hooggeëerde smaragden Boeddha huisvestte.

Chiang Mai

Wat Tha Ton

Rice fields

Buddha statues, Wat Phra That Doi Suthep

Wat Phra That Doi Suthep

This temple is one of the holiest Buddhist sites in northern Thailand. It is located on the flanks of the 1676 m high mountain Doi Suthep, and offers a magnificent view of Chiang Mai. Like most of the buildings and figures here, the over 30 m high gilded chedi (stupa) and its filigree canopies date from the 16th century.

Wat Phra That Doi Suthep

El templo es uno de los sitios budistas más sagrados del norte de Tailandia. Se encuentra en el flanco de la montaña Doi Suthep, de 1676 m de altura y ofrece una maravillosa vista de Chiang Mai. El Chedi dorado de más de 30 m de altura y sus paraguas de filigrana datan del siglo XVI, como la mayoría de los edificios y figuras.

Wat Phra That Doi Suthep

Ce temple est l'un des sites bouddhistes les plus sacrés du nord de la Thaïlande. Il se trouve sur le flanc du Doi Suthep, haut de de 1676 m, et offre une vue magnifique sur Chiang Mai. Le chedi doré de plus de 30 m de haut et ses parasols en filigrane, comme la plupart des bâtiments et des statues, datent du XVIᵉ siècle.

Wat Phra That Doi Suthep

Il tempio è uno dei luoghi più sacri del nord della Thailandia. Si trova sul fianco del monte Doi Suthep, a 1.676 m di altezza, e offre una splendida vista su Chiang Mai. Il chedi, con i suoi 30 m e oltre, e i suoi ombrelli finemente lavorati in filigrana, risalgono al XVI secolo, come la maggior parte degli edifici e delle statue qui presenti.

Wat Phra That Doi Suthep

Der Tempel gehört zu den heiligsten buddhistischen Stätten im nördlichen Thailand. Er liegt an der Bergflanke des 1676 m hohen Doi Suthep und bietet einen herrlichen Blick auf Chiang Mai. Der über 30 m hohe vergoldete Chedi und seine ihn umgebenen filigranen Schirme stammen wie die meisten Gebäude und Figuren aus dem 16. Jahrhundert.

Wat Phra That Doi Suthep

Deze tempel behoort tot de heiligste boeddhistische plaatsen van Noord-Thailand. Hij ligt op de berghelling van de 1676 meter hoge Doi Suthep en biedt een prachtig uitzicht op Chiang Mai. De meer dan 30 meter hoge vergulde chedi en de parasol van filigraanwerk ernaast dateren, net als de meeste gebouwen en beelden, uit de 16e eeuw.

Doi Chiang Dao Mountain

Chiang Rai

Phu Chi Fa Mountain

Mae Kok River

Chiang Rai

Thailand's northernmost province harmoniously combines romantic river landscapes and fascinating rock formations. In between, there are lively places such as Mae Kok, the provincial capital of Chiang Rai, and remote mountain ranges that extend to the border of Myanmar and Laos: the "Golden Triangle".

Chiang Rai

La provincia más septentrional de Tailandia combina armoniosamente románticos paisajes fluviales y fascinantes formaciones rocosas. En medio hay lugares animados como la capital de la provincia, Chiang Rai, en Mae Kok, y remotas cadenas montañosas que se extienden hasta la frontera de Myanmar y Laos, conformando el llamado "Triángulo de Oro".

Chiang Rai

La province la plus au nord de la Thaïlande est un harmonieux mélange de paysages fluviaux romantiques et de formations rocheuses fascinantes. On y trouve également des endroits animés, comme la capitale provinciale Chiang Rai, sur le Mae Kok, ainsi que des chaînes de montagnes qui s'étendent jusqu'à la frontière du Myanmar et du Laos – une région communément surnommé le « Triangle d'or ».

Chiang Rai

La provincia più settentrionale della Thailandia riunisce armoniosamente romantici paesaggi fluviali e spettacolari formazioni rocciose. La regione è costellata di città vivaci, come il capoluogo di provincia Chiang Rai lungo il Mae Kok, e remote catene montuose che si estendono fino al cosiddetto "triangolo d'oro", la zona di confine con Myanmar e Laos.

Chiang Rai

Die nördlichste Provinz Thailands vereint auf harmonische Weise romantische Flusslandschaften und faszinierende Felsformationen. Dazwischen verteilen sich lebendige Orte wie die am Mae Kok liegende Provinzhauptstadt Chiang Rai und abgelegene Gebirgszüge, die bis zur Grenze von Myanmar und Laos, dem „Goldenen Dreieck" reichen.

Chiang Rai

De noordelijkste provincie van Thailand verenigt op harmonieuze wijze romantische rivierlandschappen met fascinerende rotsformaties. Daartussen liggen levendige plaatsen zoals de aan de Mae Kok-rivier gelegen provinciehoofdstad Chiang Rai en afgelegen bergketens die zich uitstrekken tot de aan grens met Myanmar en Laos, de zogenaamde "Gouden Driehoek".

Mekong River, Chiang Khong

Near the border area to Laos

Chiang Khong is idyllically situated on the banks of the Mekong River on the border with Laos. Here the Tai Lue people live, who primarily use the river as a trade route for their goods. The few tourists who come here mainly use the town for a short excursion to Laos. Near Chiang Rai is the enchanting Wat Huay Pla Kang. The 100 m high white Guanyin statue is visible from afar, especially in the morning and evening when it is bathed in wonderful light. A nine-storey pagoda towers over the vast area, and from its upper floors there is a unique panoramic view of the mountains of northern Thailand.

Zone frontalière du Laos

Chiang Khong est idéalement situé sur les rives du Mékong, à la frontière du Laos. Ici vivent les membres du peuple Tai Lue, qui utilisent principalement le fleuve comme route commerciale pour leurs marchandises. Les quelques touristes qui parcourent la région le font surtout dans le cadre d'excursions vers le Laos. À Chiang Rai se dresse l'enchanteur Wat Huay Pla Kang et sa statue de Guanyin, déesse de la compassion, haute de 100 m et baignée d'une lumière magnifique, surtout le matin et le soir. Une pagode de neuf étages trône également sur l'immense terrain. Les étages supérieurs offrent une vue panoramique unique sur les montagnes du nord de la Thaïlande.

Im Grenzgebiet zu Laos

Chiang Khong liegt idyllisch am Ufer des Mekong an der Grenze zu Laos. Hier leben die Angehörigen vom Volk der Tai Lue, die vornehmlich den Fluss als Handelsweg für ihre Waren nutzen. Die wenigen Touristen nutzen den Ort vor allem für einen kurzen Abstecher nach Laos. Bei Chiang Rai erhebt sich der bezaubernde Wat Huay Pla Kang. Von weitem unübersehbar ist die rund 100 m hohe weiße Guanyin-Statue, die besonders am Morgen und am Abend von der Sonne in ein wunderbares Licht getaucht wird. Auf dem riesigen Areal thront ferner eine neunstöckige Pagode. Von deren oberen Etagen ergibt sich ein einmaliger Rundblick auf die Bergwelt Nordthailands.

Wat Huay Pla Kang

En la zona fronteriza con Laos

Chiang Khong se ubica en un entorno idílico, a orillas del río Mekong, en la frontera con Laos. Aquí viven los miembros del pueblo Tai Lue, que principalmente utilizan el río como ruta de comercio para sus mercancías. Los pocos turistas visitantes, parten del lugar para realizar una breve excursión a Laos. En Chiang Rai se levanta el fascinate templo de Wat Huay Pla Kang. La estatua blanca de Guanyin, de 100 m de altura, que está bañada por una luz maravillosa, especialmente por la mañana y por la tarde, se puede atisbar desde lejos. Una pagoda de nueve pisos también se encuentra entronizada en el enorme complejo. Desde los pisos superiores hay una vista panorámica única de las montañas del norte de Tailandia.

Al confine con il Laos

L'idilliaca città di Chiang Khong si trova sulle rive del fiume Mekong, al confine con il Laos. Qui vive il popolo dei Tai Lue, che utilizzano il fiume principalmente come via di commercio per le loro merci. Ai pochi turisti che qui giungono la città serve come base per una breve escursione in Laos. A Chiang Rai sorge l'incantevole Wat Huay Pla Kang. La statua del Guanyin Bianco, alta 100 metri, immersa in una luce meravigliosa soprattutto la mattina e la sera, può essere vista anche da lontano. Una pagoda di nove piani troneggia su tutta l'area. Dai piani superiori si gode di una vista panoramica straordinaria sulle montagne del nord della Thailandia.

In het grensgebied met Laos

Chiang Khong ligt idyllisch aan de oever van de Mekong op de grens met Laos. Hier wonen leden van het Tai Lue-volk, die de rivier vooral gebruiken als handelsroute voor hun goederen. De weinige toeristen doen de plaats vooral aan voor een korte excursie naar Laos. Bij Chiang Rai verrijst de betoverende Wat Huay Pla Kang. Het 100 meter hoge witte Guanyinbeeld, dat vooral 's morgens en 's avonds in een prachtig licht baadt, is van veraf al te zien. Verder staat er een pagode van negen verdiepingen op het enorme terrein. Vanaf de bovenste verdiepingen heeft u een uniek panoramisch uitzicht op de bergen van Noord-Thailand.

Natural hot springs, Lam Nam Kok National Park

Primeval forest and hot springs

The Chiang Rai region captivates with its impressive landscapes, impenetrable jungles, glittering waterfalls and lonely mountain lakes. A very special experience is a visit to the natural hot springs. The water, rich in sulphur and minerals, reaches temperatures of over 90 °C. A mystical world far from the hustle and bustle of the city.

Selva y aguas termales

La región Chiang Rai cautiva por sus impresionantes paisajes con bosques impenetrables, cascadas brillantes y lagos solitarios de montaña. La visita a las termas naturales constituye una experiencia muy especial. El agua, rica en azufre y minerales, alcanza temperaturas superiores a los 90 °C. Un paisaje místico lejos del bullicio de la ciudad.

Jungle et sources chaudes

La région de Chiang Rai séduit par ses paysages impressionnants couverts de forêts vierges impénétrables, ses cascades scintillantes et ses lacs de montagnes peu fréquentés. La visite des sources chaudes naturelles est une expérience particulière : l'eau, riche en soufre et en minéraux, y atteint des températures supérieures à 90 °C. Un paysage mystique, loin de l'agitation de la ville.

Giungla e sorgenti termali

Con i suoi paesaggi imponenti, le foreste vergini impenetrabili, le cascate scintillanti e i laghi di montagna sparsi qua e là, la regione di Chiang Rai è una meta ricca di fascino. Un'esperienza del tutto speciale è la visita alle sorgenti termali naturali: l'acqua, ricca di zolfo e minerali, raggiunge temperature superiori ai 90 °C. Uno scenario mistico lontano dal trambusto della città.

Urwald und heiße Quellen

Die Region Chiang Rai besticht durch eindrucksvolle Landschaften mit undurchdringlichen Urwäldern, glitzernden Wasserfällen und einsamen Bergseen. Ein ganz besonderes Erlebnis ist der Besuch der natürlichen heißen Quellen. Das an Schwefel und Mineralien reiche Wasser erreicht Temperaturen bis über 90 °C. Eine mystische Szenerie fernab des städtischen Trubels.

Oerwoud en warmwaterbronnen

De regio Chiang Rai bekoort door indrukwekkende landschappen met ondoordringbare oerwouden, schitterende watervallen en eenzame bergmeren. Een bijzondere belevenis is een bezoek aan de natuurlijke warmwaterbronnen. Het zwavel- en mineraalrijke water heeft een temperatuur van ruim 90 °C. Een mystiek landschap ver weg van de drukte van de stad.

Lam Nam Kok National Park

Wat Rong Khun

Not all temples in Thailand can look back on centuries
of history. Construction of this site in Chiang Rai,
also known as the "White Temple", began in 1997
and just a fraction of the planned complex has been
completed to date. The extravagant design is by the
Thai architect and artist Chalermchai Kositpipat.

Wat Rong Khun

Tous les temples de Thaïlande ne sont pas les témoins
de siècles d'histoire. La construction du site de Chiang
Rai, également connu sous le nom de « Temple
blanc », n'a débuté qu'en 1997, et seule une partie du
complexe prévu a été achevée à ce jour. Le design,
extravagant, a été imaginé par l'architecte et artiste
thaïlandais Chalermchai Kositpipat.

Wat Rong Khun

Nicht alle Tempel in Thailand können auf eine
jahrhundertealte Geschichte zurückblicken. Mit dem
Bau, der auch als „Weißer Tempel" bekannten Stätte
in Chiang Rai begann man erst im Jahr 1997. Bis heute
ist nur ein Bruchteil der geplanten Gesamtanlage
fertiggestellt. Der extravagante Designentwurf
stammt vom thailändischen Architekten und Künstler
Chalermchai Kositpipat.

Wat Rong Khun

No todos los templos de Tailandia pueden recordar
siglos de historia. La construcción del complejo en
Chiang Rai, también conocido como el "Templo
Blanco", no comenzó hasta 1997 y sólo se ha
completado una fracción del mismo hasta la fecha. El
extravagante diseño fue obra del arquitecto y artista
tailandés Chalermchai Kositpipat.

Wat Rong Khun

Non tutti i templi della Thailandia hanno una storia
di secoli. La costruzione del sito di Chiang Rai,
noto anche come il "tempio bianco", è stata avviata
nel 1997 e solo una parte del complesso è stata
finora completata. Il design stravagante è stato
progettato dall'architetto e artista thailandese
Chalermchai Kositpipat.

Wat Rong Khun

Niet alle tempels in Thailand kunnen terugkijken op
een eeuwenoude geschiedenis. De bouw van het ook
wel als de "Witte tempel" bekendstaande complex in
Chiang Rai begon pas in 1997. Tot nog toe is slechts
een fractie van het geplande complex voltooid. Het
extravagante ontwerp is van de Thaise architect en
kunstenaar Chalermchai Kositpipat.

Golden Triangle at Mekong River

PEE TA KHON, LOEI

WAT KHAO PHRA KHRU, CHONBURI

FLOWER GARLAND

ROYAL BARGE PROCESSION, BANGKOK

HOLY TREE

SHADOW PLAY

LOI KRATHONG

WAT JED YOD, CHIANG MAI

SONGKRAN

OFFERING

LOI KRATHONG, BANGKOK

FLOATING FLOWERS

OFFERING

TRADITIONS

Throughout the year the Thais celebrate traditional festivals with great devotion. Unforgettable moments are offered by "Loi Krathong", the Festival of Lights, or "Songkran", the New Year Festival. The decoration and honoring of sacred sites, whether trees or temples, is a matter of course for the people, whose lives are predominantly deeply rooted in Buddhism and religion.

TRADITIONS

Tout au long de l'année, les Thaïlandais célèbrent avec dévotion les fêtes traditionnelles. Le festival des lanternes « Loi Krathong » ou le festival du nouvel an « Songkran » offrent ainsi des moments inoubliables. Décorer et honorer des sites sacrés, qu'il s'agisse d'arbres ou de temples, est une évidence dans la vie des habitants qui sont pour la plupart profondément attachés au bouddhisme et à la religion.

BRAUCHTUM

Rund um das Jahr feiern die Thailänder mit Hingabe traditionelle Feste. Unvergessliche Momente offeriert das Lichterfest „Loi Krathong" oder das Neujahrfest „Songkran". Das Schmücken und Ehren von heiligen Stätten, ob Baum oder Tempel, ist eine Selbstverständlichkeit im Leben der meist buddhistischen und religiös tiefverwurzelten Menschen.

COSTUMBRES

Durante todo el año, los tailandeses celebran con devoción las fiestas tradicionales. el Festival de las Luces "Loi Krathong" o el Festival de Año Nuevo "Songkran" ofrecen momentos inolvidables .Decorar y honrar los lugares sagrados, ya sean árboles o templos, es algo natural en la vida de las personas que en su mayoría están profundamente arraigadas en el budismo y la religión.

TRADIZIONI

Durante tutto l'anno i thailandesi celebrano feste tradizionali molto sentite. Momenti indimenticabili sono offerti dal Festival delle luci, "Loi Krathong", o dal Festival di Capodanno, "Songkran". Decorare e venerare i luoghi sacri, siano essi alberi o templi, fa parte della vita quotidiana per i buddisti, che vivono molto intensamente l'aspetto spirituale.

GEBRUIKEN

De Thai vieren het hele jaar door met toewijding traditionele feesten. Onvergetelijke momenten biedt het lichtfeest "Loi Krathong" of het Nieuwjaarsfeest "Songkran". Het versieren en eren van heilige plaatsen, of het nu bomen of tempels zijn, is een vanzelfsprekendheid in het leven van mensen die meestal sterk boeddhistisch en religieus zijn.

Mae Hong Son, Tak & Lamphun

Pai Canyon, Mae Hong Son

Mae Hong Son

Mae Hong Son, Tak & Lamphun

Jungle, rivers and mountains as far as the eye can see. The remote province of Mae Hong Son, bordering Myanmar with the 1685 m high Doi Pui surprises with few tourists and the charming Namtok Mae Surin National Park; perfect for a trekking tour. Further south is equally sparsely populated Tak province with the city of Mae Sot, which boasts a number of Buddhist temples such as Wat Manee Pai Son. Buddhist sanctuaries of great national importance are located in the old town of Lamphun. Visitors are lured to the 1000 km² Mae Ping National Park, where bears still live.

Mae Hong Son, Tak & Lamphun

Jungle, rivières et montagnes à perte de vue – parmi lesquelles le Doi Pui, de 1685 m de haut – ; la province éloignée de Mae Hong Son, bordant le Myanmar, est pourtant très peu touristique, mais le charmant parc national de Namtok Mae Surin est parfait pour un trekking. Plus au sud, la province de Tak est également peu peuplée ; on y trouve la ville de Mae Sot et des temples bouddhistes comme Wat Manee Pai Son. Les sanctuaires d'importance nationale sont situés dans la vieille ville de Lamphun, dans la province voisine. À cheval sur plusieurs de ces régions, le parc national de Mae Ping, d'une superficie d'environ 1 000 km², est l'habitat de plusieurs ours.

Mae Hong Son, Tak & Lamphun

Dschungel, Flüsse und Berge soweit das Auge reicht. Die abgelegene, an Myanmar angrenzende Provinz Mae Hong Son mit dem 1685 m hohen Doi Pui überrascht mit wenig Touristen, dem reizvollen Namtok Mae Surin Nationalpark und ist perfekt für eine Trekking-Tour geeignet. Weiter südlich erstreckt sich die ebenfalls dünn besiedelte Provinz Tak mit der Stadt Mae Sot, die durch sehenswerte buddhistische Tempel wie den Wat Manee Pai Son begeistert. Buddhistische Heiligtümer von hoher nationaler Bedeutung präsentieren sich in der Altstadt von Lamphun. In der Umgebung lockt der rund 1000 km² große Nationalpark Mae Ping, in dem noch einige Bären leben.

Tak

Mae Hong Son, Tak & Lamphun

Selva, ríos y montañas hasta donde alcanza la vista. La remota provincia de Mae Hong Son, que linda con Myanmar, y que cuenta con el Doi Pui de 1685 m de altura, sorprende por los pocos turistas que la visitan. El fascinante parque nacional de Namtok Mae Surin es perfecto para una excursión de senderismo. Más al sur se encuentrala provincia de Tak, también escasamente poblada, con la ciudad de Mae Sot, que cautiva con sus templos budistas como el Wat Manee Pai Son. En el casco antiguo de Lamphun se encuentran santuarios budistas de gran importancia nacional. El Parque nacional de Mae Ping, de unos 1000 km² de superficie, atrae a algunos osos.

Mae Hong Son, Tak & Lamphun

Giungla, fiumi e montagne a perdita d'occhio: al confine con il Myanmar, la remota provincia di Mae Hong Son, su cui svetta il Doi Pui con i suoi 1.685 m di altezza, sorprende per i pochi turisti. Il Parco nazionale di Namtok Mae Surin è una meta ideale per il trekking. Più a sud si trova la provincia di Tak, anch'essa scarsamente popolata. Qui la città di Mae Sot offre uno spettacolo affascinante con i suoi vari templi buddisti, come quello di Wat Manee Pai Son. Nel centro storico di Lamphun si trovano santuari buddisti di rilevanza nazionale. Nel Parco nazionale di Mae Ping, con una superficie di circa 1.000 km², meta molto apprezzata, vivono ancora alcuni orsi.

Mae Hong Son, Tak & Lamphun

Jungle, rivieren en bergen zo ver als het oog reikt. De afgelegen, aan Myanmar grenzende provincie Mae Hong Son, met de 1685 meter hoge berg Doi Pui, verrast met weinig toeristen en het charmante nationale park Namtok Mae Surin en leent zich perfect voor een lange wandeltocht. Verder naar het zuiden ligt de eveneens dunbevolkte provincie Tak met de stad Mae Sot, die fascineert met zijn boeddhistische tempels zoals de Wat Manee Pai Son. Boeddhistische heiligdommen van groot nationaal belang staan in het oude centrum van Lamphun. In de omgeving lokt het ongeveer 1000 km² grote nationale park Mae Ping, waar nog enkele beren leven.

Mae Hong Son

Mae Ping National Park, Lamphun

Wat Phra That Doi Din Chi, Mae Sot, Tak

Mae Sot

The lively and bustling border town of Mae Sot in Tak province is greatly influenced by the neighbouring country of Myanmar, both economically and culturally. The hill tribes living in the region have left many lasting traces: Karen people created the pagoda Wat Phra That Doi Din Chi, which stands on a narrow golden rock. In the centre of Mae Sot is Wat Manee Pai Son, a spacious religious complex of considerable beauty. The huge Samphutte-Chedi rises impressively, framed by 233 smaller chedis. In each one there is a portrait of Buddha.

Mae Sot

La ville animée de Mae Sot, dans la province de Tak, est, tant sur le plan économique que culturel, très proche du pays voisin, le Myanmar. Les montagnards de la région y ont durablement laissé leur empreinte, comme le prouve la pagode Wat Phra That Doi Din Chi, perchée sur un étroit rocher doré, qui a été construite par le peuple Karen. Au centre de Mae Sot se trouve le Wat Manee Pai Son, un vaste complexe religieux d'une beauté considérable. L'immense chedi « Samphutte », encadré par 233 plus petits chedis – qui contiennent chacun un portrait de Bouddha – est particulièrement impressionnant.

Mae Sot

Die quirlige und geschäftige Grenzstadt Mae Sot in der Provinz Tak ist wirtschaftlich und auch kulturell sehr stark durch die Nähe zum Nachbarland Myanmar geprägt. Die in der Region lebenden Bergvölker hinterließen zahlreiche bleibende Spuren. So schufen Angehörige vom Volk der Karen die auf einem schmalen goldenen Felsen stehende Pagode Wat Phra That Doi Din Chi. Im Zentrum vom Mae Sot zeigt sich der Wat Manee Pai Son. Ein großzügig angelegter religiöser Komplex von beachtlicher Schönheit. Imposant erhebt sich der riesige Samphutte-Chedi, der von 233 kleineren Chedis umrahmt wird. In jedem einzelnen ist ein Bildnis von Buddha untergebracht.

Wat Manee Pai Son, Mae Sot, Tak

Mae Sot

La animada y bulliciosa ciudad fronteriza de Mae Sot, en la provincia de Tak, está muy cerca del país vecino de Myanmar, tanto en materia económica como cultural. La gente de las montañas que vive en la región dejó muchos rastros que aún perduran. Así, la gente de Karen construyó la pagoda Wat Phra That Doi Din Chi sobre un estrecha montículo dorado. En el centro de Mae Sot se ubica el complejo religioso de Wat Manee Pai Son, de considerable belleza. El enorme Chedi Samphutte se eleva majestuoso, enmarcado por 233 chedis de tamaño inferior. En cada uno hay un retrato de Buda.

Mae Sot

Sia dal punto di vista economico che culturale, la vivace e indaffarata città di Mae Sot, nella provincia di Tak, è molto vicina al Myanmar, da cui dista poco. I popoli montani di queste zone hanno lasciato segni indelebili della loro esistenza. Il popolo dei Karen, ad esempio, ha edificato la pagoda di Wat Phra That Doi Din Chi, che si erge su una stretta roccia d'oro. Nel centro di Mae Sot si trova il Wat Manee Pai Son, un ampio complesso sacro di notevole bellezza. Il gigantesco chedi Samphutte si erge impressionante tra altri 233 chedi più piccoli. In ognuno di essi è collocato un ritratto del Buddha.

Mae Sot

De levendige en bedrijvige grensplaats Mae Sot in de provincie Tak is zowel economisch als cultureel sterk beïnvloed door het nabijgelegen buurland Myanmar. De bergvolken uit de regio hebben veel blijvende sporen achtergelaten. Zo bouwden leden van het Karen-volk de pagode Wat Phrathat Doi Din Chi op een smalle gouden rots. In het centrum van Mae Sot staat Wat Manee Pai Son, een groots aangelegd religieus complex van aanzienlijke schoonheid. Imposant verheft de enorme Samphutte-chedi zich, omgeven door 233 kleinere chedi's. In elke chedi staat een Boeddhabeeld.

Nam Tok Pha Charoen National Park, Tak

Tulay Hill, Tak

Huai Kha Khaeng Wildlife Sanctuary

Huai Kha Khaeng Wildlife Sanctuary

This 3000 km² game reserve was included in the UNESCO World Heritage List in 1991. Some areas of the mostly mountainous and forest-covered reserve lie in Tak province, the rest in Uthai Thani and Kanchanaburi provinces. With a bit of luck, elephants, leopards and deer can be seen on an extended tour. Although the largest tiger population of Thailand makes its home here, these majestic animals are very rare to see. Banteng, a species of cattle native to Southeast Asia, or wild water buffalo are much more common.

Réserve de Huai Kha Khaeng

Cette réserve de 3000 km² a été inscrite au patrimoine mondial de l'UNESCO en 1991. Certaines zones, essentiellement montagneuses et boisées, se trouvent dans la province de Tak, le reste appartient aux provinces d'Uthai Thani et de Kanchanaburi. Avec un peu de chance, on peut y apercevoir des éléphants, des léopards et des cerfs ; il est plus rare d'y rencontrer un tigre, bien que la réserve soit le lieu de résidence de la plus grande population du majestueux animal en Thaïlande. Les bantengs, des vaches sauvages originaires d'Asie du Sud-Est, ou les buffles d'eau sauvages y sont beaucoup plus courants.

Huai Kha Khaeng Widreservat

Das knapp 3000 km² umfassende Wildreservat wurde 1991 in die UNESCO-Liste des Weltkulturerbes aufgenommen. Einige Bereiche des meist bergigen und von Wald bedeckten Schutzgebietes liegen in der Tak-Provinz, der Rest in den Provinzen Uthai Thani und Kanchanaburi. Mit etwas Glück können bei einer ausgedehnten Tour Elefanten, Leoparden und Hirsche gesichtet werden. Obwohl hier die größte Tiger-Population Thailands lebt und zuhause ist, sind diese majestätischen Tiere sehr selten aufzuspüren. Deutlich häufiger erblickt man ein Banteng, ein in Südostasien beheimatetes Wildrind, oder wilde Wasserbüffel.

Huai Kha Khaeng Wildlife Sanctuary

Coto de caza Huai Kha Khaeng

La reserva de 3000 km² fue incluida en la Lista del
Patrimonio Mundial de la UNESCO en 1991. Algunas
zonas de la reserva, en su mayoría montañosas y
boscosas, se encuentran en la provincia de Tak, el
resto en las provincias de Uthai Thani y Kanchanaburi.
Con un poco de suerte, se pueden ver elefantes,
leopardos y ciervos en un tour ampliado. A pesar de
que la mayor población de tigres de Tailandia vive y
tiene su higar aquí, estos majestuosos animales son
muy difíciles de encontrar. Un *banteng,* una vaca
salvaje autóctona del sudeste asiático, o un búfalo de
agua salvaje son mucho más comunes.

Riserva faunistica di Huai Kha Khaeng

Questa riserva faunistica, di poco meno di 3.000 km²,
è stata inserita nella lista del patrimonio dell'UNESCO
nel 1991. Alcune zone della riserva prevalentemente
montuosa e boschiva fanno capo alla provincia di
Tak, mentre le altre si trovano nelle province di Uthai
Thani e Kanchanaburi. Con un po' di fortuna, elefanti,
leopardi e cervi possono essere ammirati durante i
tour più lunghi. Anche se la più grande popolazione di
tigri della Thailandia vive qui, questi animali maestosi
vengono avvistati solo molto raramente. Molto più
frequente è invece il banteng, o bufalo d'acqua, un
bovino selvatico originario del sud-est asiatico.

Wildreservaat Huai Kha Khaeng

Het wildreservaat van bijna 3000 km² stond in 1991
op de werelderfgoedlijst van de Unesco. Enkele
delen van het voornamelijk bergachtige en met bos
bedekte reservaat liggen in de provincie Tak, de rest
in de provincies Uthai Thani en Kanchanaburi. Met een
beetje geluk zijn olifanten, luipaarden en herten te
zien tijdens een uitgebreide tour. Hoewel de grootste
tijgerpopulatie van Thailand hier leeft en thuis is,
zijn deze majestueuze dieren zelden te zien. Een
banteng, een wild rund uit Zuidoost-Azië, of een wilde
waterbuffel komt veel vaker voor.

Phayao, Nan & Phrae

Doi Phu Kha National Park, Nan

Wat Phumin, Nan

Phayao, Nan & Phrae

These three provinces are located in the north-east of Thailand, and are some to the most beautiful regions that the north hides in its jewellery box. From the top of Phu Lang Ka, in the national park of the same name, the view over the misty landscape of Phayao province is a feast for the eyes. The city of Nan promises cultural highlights. This gem has Wat Phumin, a spectacular temple complex built in 1596. The capital of the neighbouring province Phrae is even older. It was founded on the Yom River in the 11th century. The impressive city wall around the old centre dates back to this period.

Phayao, Nan & Phrae

Ces trois provinces sont situées au nord-est de la Thaïlande et font partie des plus belles régions de la zone. Du haut du Phu Lang Ka, dans le parc national du même nom, la vue sur le paysage brumeux de la province de Phayao est un régal pour les yeux. La ville de Nan propose de nombreux divertissements culturels. Mais le plus grand joyau est peut-être le Wat Phumin, un spectaculaire complexe de temples construit en 1596. La capitale de la province voisine Phrae, sur la rivière Yom, est encore plus ancienne : elle a été fondée au XIe siècle. Datant de cette même période, la muraille de la ville, autour du vieux centre, est impressionnante.

Phayao, Nan & Phrae

Die drei Provinzen liegen im Nordosten Thailands und gehören zu den schönsten Regionen, die der Norden in seinem Schmuckkästchen versteckt. Vom Gipfel des Phu Lang Ka, im gleichnamigen Nationalpark, ist die Aussicht über die nebelverhangene Landschaft der Provinz Phayao eine Augenweide. Kulturelle Höhepunkte verspricht die Stadt Nan. Das Kleinod besitzt den Wat Phumin, eine spektakuläre, im Jahr 1596 erbaute Tempelanlage. Noch älter ist die Hauptstadt der Nachbarprovinz Phrae. Sie wurde im 11. Jahrhundert am Fluss Yom gegründet. Eindrucksvoll ist die, noch aus dieser Zeit stammende Stadtmauer um das alte Zentrum.

Phu Langka Forest Park, Phayao

Phayao, Nan & Phrae

Las tres provincias están situadas en el noreste de Tailandia y pertenecen a las regiones más bellas que el norte esconde en su joyero. Desde la cima de Phu Lang Ka, en el parque nacional del mismo nombre, la vista sobre el paisaje nebuloso de la provincia de Phayao es un regalo para la vista. La ciudad de Nan promete atractivos culturalescomo el espectacular complejo Wat Phumin, construido en 1596. La capital de la provincia vecina Phrae es aún más antigua, se fundó en el siglo XI sobre el río Yom. La muralla que rodea el casco antiguo, que data de esa época, es impresionante.

Phayao, Nan & Phrae

Le tre province si trovano nel nord-est della Thailandia e sono tra le regioni più belle del nord. Dalla cima del Phu Lang Ka, nell'omonimo parco nazionale, la vista sul paesaggio nebbioso della provincia di Phayao è uno spettacolo da non perdere. La città di Nan è invece conosciuta per i siti di importanza culturale. Questo piccolo gioiello racchiude infatti il Wat Phumin, uno spettacolare complesso di templi costruito nel 1596. Ancora più antica è invece la capitale della vicina provincia di Phrae, fondata nel XI secolo lungo il fiume Yom, il cui centro storico, risalente allo stesso periodo, è circondato da un'imponente cinta muraria.

Phayao, Nan & Phrae

Deze drie provincies liggen in het noordoosten van Thailand en behoren tot de mooiste regio's die het noorden in zijn sieradenkistje heeft verstopt. Vanaf de top van de Phu Lang Ka, in het gelijknamige nationale park, is het uitzicht over het mistige landschap van de provincie Phayao een lust voor het oog. De stad Nan belooft culturele hoogtepunten. Deze schitterende stad bezit het spectaculaire tempelcomplex Wat Phumin, dat in 1596 werd gebouwd. De hoofdstad van de aangrenzende provincie Phrae is nog ouder. Hij werd in de 11e eeuw gesticht aan de rivier de Yom. De stadsmuur rond het oude centrum, die nog uit deze tijd stamt, is erg indrukwekkend.

Phu Langka Forest Park, Phayao

Phayao Lake, Phayao

Phayao Lake
Phayao Lake, beautifully embedded in nature, is the largest freshwater lake in northern Thailand with an area of over 2 km² and is the main attraction of the area. The average water depth of just under 2 m provides a habitat for more than 45 species of fish. Take a taxi boat from the shore if you want to make a leisurely tour.

Lac Phayao
Le lac Phayao est, avec une superficie de plus de 2 km², le plus grand lac d'eau douce du nord de la Thaïlande. C'est l'attraction principale de la région. Sa profondeur, d'en moyenne moins de 2 mètres, en fait un habitat idéal pour plus de 45 espèces de poissons. Pour une excursion tranquille, prenez un bateau-taxi sur le rivage.

Phayao-See
Der wunderschön in die Natur eingebettete Phayao-See ist mit einer Fläche von gut 2 km² der größte Süßwassersee Nordthailands. Er ist die Hauptattraktion der Gegend. Das durchschnittlich nur knapp 2 m Tiefe Gewässer ist Lebensraum für mehr als 45 Fischarten. Wer eine geruhsame Tour unternehmen möchte, nimmt sich am Ufer ein Taxiboot.

Phayao Lake, Phayao

Lago Phayao

El lago Phayao, está enclavado en la naturaleza y es la reservade agua dulce más grande del norte de Tailandia con una superficie de más de 2 km². Constituye la principal atracción de la zona. La profundidad media del agua de poco menos de 2 m es el hábitat de más de 45 especies de peces. Si quiere hacer una excursión tranquila, tome un bote taxi en la orilla.

Lago di Phayao

Il lago di Phayao, splendidamente incastonato nella natura circostante, con una superficie di oltre 2 km² è il più grande lago d'acqua dolce del nord della Thailandia e la principale attrazione della zona. Le sue acque, profonde poco meno di 2 m, sono l'habitat di più di 45 specie ittiche. Per chi preferisce escursioni non faticose è consigliabile un taxi-boat.

Phayaomeer

Het prachtig in de natuur ingebedde Phayaomeer is met een oppervlakte van meer dan 2 km² het grootste zoetwatermeer van Noord-Thailand. Het is de belangrijkste attractie van het gebied. Het gemiddeld slechts iets minder dan 2 meter diepe water vormt een habitat voor meer dan 45 vissoorten. Als u een rustige tour wilt maken, neem dan een taxiboot aan de oever.

Pua District, Nan

Phae Mueang Phi Forest Park, Phrae

Feet of Buddha, Wat Phra That Suthon Mongkol Khiri, Phrae

Wat Phra That Suthon Mongkol Khiri

The Buddhist temple Wat Phra That Suthon Mongkol Khiri near the city of Phrae is not a very old sanctuary, and is constantly being extended with new buildings. Especially fascinating are the different stylistic elements and the recently created 89 m long reclining Buddha figure.

Wat Phra That Suthon Mongkol Khiri

Le temple bouddhiste Wat Phra That Suthon Mongkol Khiri, près de la ville de Phrae, n'est pas un sanctuaire très ancien, mais il est constamment agrandi avec de nouveaux bâtiments. Les différents éléments stylistiques, mais aussi le Bouddha couché, long de 89 mètres, sont particulièrement impressionnants.

Wat Phra That Suthon Mongkol Khiri

Die buddhistische Tempelanlage Wat Phra That Suthon Mongkol Khiri nahe der Stadt Phrae ist kein sehr altes Heiligtum. Immer wieder wird sie durch neue Bauten erweitert. Besonders faszinierend sind die verschiedenen Stilelemente und die nur wenige Jahre alte, liegende Buddhafigur, mit einer Länge von stolzen 89 m.

Wat Phra That Suthon Mongkol Khiri, Phrae

Wat Phra That Suthon Mongkol Khiri

El templo budista Wat Phra That Suthon Mongkol Khiri cerca de la ciudad de Phrae no es un santuario muy antiguo y se está ampliando constantemente con nuevos edificios. Los diferentes elementos estilísticos y la única figura de Buda de unos pocos años, con una longitud de 89 m despiertan una fascinación especial.

Wat Phra That Suthon Mongkol Khiri

Il tempio buddista di Wat Phra That Suthon Mongkol Khiri, vicino alla città di Phrae, è piuttosto recente. Il sito viene costantemente ampliato con nuovi edifici. Notevoli i diversi elementi stilistici e la statua del Buddha disteso, realizzata solo pochi anni fa, con una lunghezza di ben 89 m.

Wat Phra That Suthon Mongkol Khiri

De boeddhistische tempel Wat Phra That Suthon Mongkol Khiri in de buurt van de stad Phrae is geen erg oud heiligdom. Het wordt voortdurend uitgebreid met nieuwe gebouwen. Bijzonder fascinerend zijn de verschillende stijlelementen en de slechts enkele jaren oude, liggende Boeddha van 89 meter lang.

Lampang

Lom Phu Kiew

Wat Chalermprakiat Prajomklao Rachanusorn

Lampang

The province of Lampang, which extends far north, offers a wide variety of sacred sites. Some of the partly filigreed temples can be discovered in the city of Lampang, others are built on secluded and rugged rocks. Beautiful lakes and picturesque waterfalls such as the Wang Kaew Waterfall are other attractions in the region.

Lampang

La province de Lampang, qui s'étend loin au nord du pays, offre une grande variété de sites sacrés. Certains des temples, partiellement en filigrane, peuvent être découverts dans la ville de Lampang, tandis que d'autres ont été construits à distance, sur un terrain rocheux accidenté. Des lacs charmants et des cascades pittoresques, comme celle de Wang Kaew, s'ajoutent aux attractions de la région.

Lampang

Die sich weit nördlich ausbreitende Provinz Lampang offeriert mannigfaltige heilige Stätten. Einige der teils filigran gestalteten Tempel lassen sich in der Stadt Lampang entdecken, andere erbaute man abgelegen auf schroffem Felsuntergrund. Anmutige Seen und malerische Wasserfälle wie der Wang Kaew Wasserfall sind weitere Attraktionen der Region.

Lampang

La provincia de Lampang, que se extiende hacia el norte, ofrece una gran variedad de lugares sagrados. Algunos de los templos, en parte filigranados, se pueden descubrir en la ciudad de Lampang, otros se construyeron alejados, en terrenos rocosos y escarpados. Lagos graciosos y pintorescas cascadas como la Wang Kaew Waterfall forman parte de las atracciones de la región.

Lampang

La provincia di Lampang, nell'estremo nord, offre una grande varietà di luoghi sacri. Diversi templi, in parte realizzati in filigrana, si trovano nella città di Lampang, mentre altri sono stati costruiti in zone remote su ripidi pendii. Laghi e cascate pittoresche, come quella di Wang Kaew, sono altre attrazioni della regione.

Lampang

De zich tot ver in het noorden uitstrekkende provincie Lampang biedt een grote verscheidenheid aan heilige plaatsen. Enkele van de deels met filigraan versierde tempels zijn te bewonderen in de stad Lampang, andere zijn op afgelegen plaatsen gebouwd op een kale rotsgrond. Lieflijke meren en schilderachtige watervallen zoals de Wang Kaew zijn andere attracties in de regio.

Spirit houses, Mae Tha

Wang Kaew Waterfall

Wat Phaphutthabat Phuphadaeng

Hot springs, Chae Son National Park

SPIRIT HOUSES

The spirit houses known in Thailand as *San Phra Phum* can be found wherever a new building has been erected, because with the building of a house or a street out the spirits originally lived at this place are driven out. To appease them, they receive a new home, usually in the form of a typical Thai temple.

MAISONS DES ESPRITS

Les maisons des esprits, appelées *San Phra Phum* en Thaïlande, se trouvent partout où un nouveau bâtiment a été érigé. Avec la construction d'une maison ou d'une rue, on chasse en effet les esprits qui vivaient à l'origine à cet endroit. Pour les apaiser, on leur installe alors une nouvelle habitation, généralement sous la forme d'un temple thaïlandais typique.

GEISTERHÄUSCHEN

Die in Thailand *San Phra Phum* genannten Geisterhäuschen sind überall dort anzutreffen, wo ein neues Gebäude errichtet wurde. Denn mit dem Bau eines Hauses oder einer Straße vertreibt man die ursprünglich an diesem Ort gelebten Geister. Um sie milde zu stimmen, erhalten sie eine neue Herberge, meist in Form eines typisch thailändischen Tempels.

CASAS FANTASMAS

Las casas fantasmas, llamadas *San Phra Phum* en Tailandia, se pueden encontrar dondequiera que se haya construido un nuevo edificio. Porque con la construcción de una casa o de una calle se expulsan los espíritus que originalmente vivían en ese lugar. Para apaciguarlos, reciben un nuevo albergue, generalmente en forma de un templo típico tailandés.

CASE DEGLI SPIRITI

In Thailandia, i *san phra phum,* le cosiddette "case degli spiriti", si trovano ovunque siano stati intrapresi lavori di edilizia. Con la costruzione di una casa o di una strada, infatti, si ritiene che vengano scacciati gli spiriti che originariamente vivevano in quel luogo. Per tenerseli amici, essi ricevono una nuova dimora, di solito a forma di tipico tempio thailandese.

GEESTENHUIZEN

De in Thailand *san phra phum* genoemde geestenhuizen zijn overal te vinden waar een nieuw gebouw wordt neergezet. Want met de bouw van een huis of de aanleg van een straat verdrijft men de geesten die oorspronkelijk op deze plaats woonden. Om ze mild te stemmen krijgen ze een nieuw onderdak, meestal in de vorm van een typisch Thaise tempel.

Sukhothai

Wat Mahathat, Sukhothai Historical Park

Wat Mahathat, Sukhothai Historical Park

Wat Chang Lom, Si Satchanalai Historical Park

Sukhothai

The historical part of Sukhothai, which was declared a UNESCO World Heritage Site in 1991, is an important national holy site. Here in the centre of Thailand, the first kingdom was founded in the 13th century. The historic park, which is worth seeing and covers several square kilometres, includes dozens of Buddha figures and temples from various eras.

Sukhothai

La parte histórica de Sukhothai, que fue declarada Patrimonio de la Humanidad por la UNESCO en 1991, es un importante santuario nacional. Aquí, en el centro de Tailandia, se fundó el primer reino del país en el siglo XIII. El parque histórico, que vale la pena visitar y cubre varios kilómetros cuadrados, incluye docenas de figuras de Buda y templos de diferentes siglos.

Sukhothai

La vieille ville de Sukhothai, qui a été inscrite au patrimoine mondial de l'UNESCO en 1991, est un important sanctuaire national : c'est ici, au centre du pays, que le premier royaume thaïlandais a été fondé au XIIIᵉ siècle. Le parc historique, qui vaut la peine d'être visité et couvre plusieurs kilomètres carrés, comprend des dizaines de figures de Bouddha et de temples de différents siècles.

Sukhothai

La parte storica di Sukhothai, dichiarata nel 1991 patrimonio dell'umanità dall'UNESCO, è un santuario di rilevanza nazionale. Qui, nel centro della Thailandia, venne fondato nel XIII secolo il primo regno del paese. Nel parco storico, meta obbligatoria che si estende per diversi chilometri quadrati, si trovano decine di figure del Buddha e templi risalenti a secoli diversi.

Sukhothai

Der historische Teil Sukhothais, der 1991 zum UNESCO-Welterbe gekürt wurde, ist ein bedeutendes nationales Heiligtum. Hier im Zentrum Thailands gründete man im 13. Jahrhundert das erste Königreich des Landes. Der sehenswerte, mehrere Quadratkilometer große Geschichtspark umfasst dutzende Buddhafiguren und Tempel aus verschiedenen Jahrhunderten.

Sukhothai

Het historische gedeelte van Sukhothai, dat in 1991 tot werelderfgoed van de Unesco werd uitgeroepen, is een heiligdom van nationaal belang. Hier in het midden van Thailand werd in de 13e eeuw het eerste koninkrijk van het land gesticht. Het historische park, dat de moeite van een bezoek waard is en meerdere vierkante kilometers beslaat, omvat tientallen Boeddhabeelden en tempels uit verschillende eeuwen.

Wat Mahathat, Sukhothai Historical Park

Sukhothai Historical Park

Wat Si Chum, Sukhothai Historical Park

Wat Si Chum

The buildings of Wat Si Chum were erected in the 14th century. The main attraction is the 15 m high seated Buddha, the largest of its kind in Sukhothai. Wat Mahathat, which is also part of the history park, is the centre of the entire area. For centuries it was the religious centre of the then Siamese kingdom.

Wat Si Chum

Les bâtiments de Wat Si Chum ont été construits au XIVe siècle. L'attraction principale est le Bouddha de 15 m de haut, assis – le plus grand de son genre à Sukhothai. Le Wat Mahathat, temple principal du parc historique, a été pendant des siècles le centre religieux du royaume siamois de l'époque.

Wat Si Chum

Die Gebäude des Wat Si Chum errichtete man im 14. Jahrhundert. Die Hauptattraktion ist der rund 15 m hohe, sitzende Buddha, der damit als größter seiner Art in Sukhothai gilt. Der ebenfalls zum Geschichtspark gehörende Wat Mahathat stellt den Mittelpunkt des gesamten Areals dar. Er war Jahrhunderte das religiöse Zentrum des damaligen siamesischen Königreichs.

Wat Mahathat, Sukhothai Historical Park

Wat Si Chum

Los edificios de Wat Si Chum fueron construidos en el siglo XIV. La atracción principal es el Buda sentado de 15 m de altura, el más grande de su clase en Sukhothai. El templo Wat Mahathat, que también pertenece al parque histórico, es el centro de toda la zona. Durante siglos fue el centro religioso del entonces reino siamés.

Wat Si Chum

Gli edifici di Wat Si Chum sono stati costruiti nel XIV secolo. L'attrazione principale è il Buddha seduto, di 15 m di altezza, il più grande del suo genere a Sukhothai. Il Wat Mahathat, anch'esso parte del parco storico, costituisce il punto saliente di tutta l'area: per secoli è stato il centro religioso dell'allora vigente regno siamese.

Wat Si Chum

De gebouwen van het tempelcomplex Wat Si Chum werden gebouwd in de 14e eeuw. De hoofdbezienswaardigheid is de 15 meter hoge, zittende Boeddha, die daarmee de grootste in zijn soort in Sukhothai is. Wat Mahathat, dat ook deel uitmaakt van het historische park, is het middelpunt van het hele terrein. Eeuwenlang was dit hét religieuze centrum van het destijds Siamese koninkrijk.

Wat Mahathat, Sukhothai Historical Park

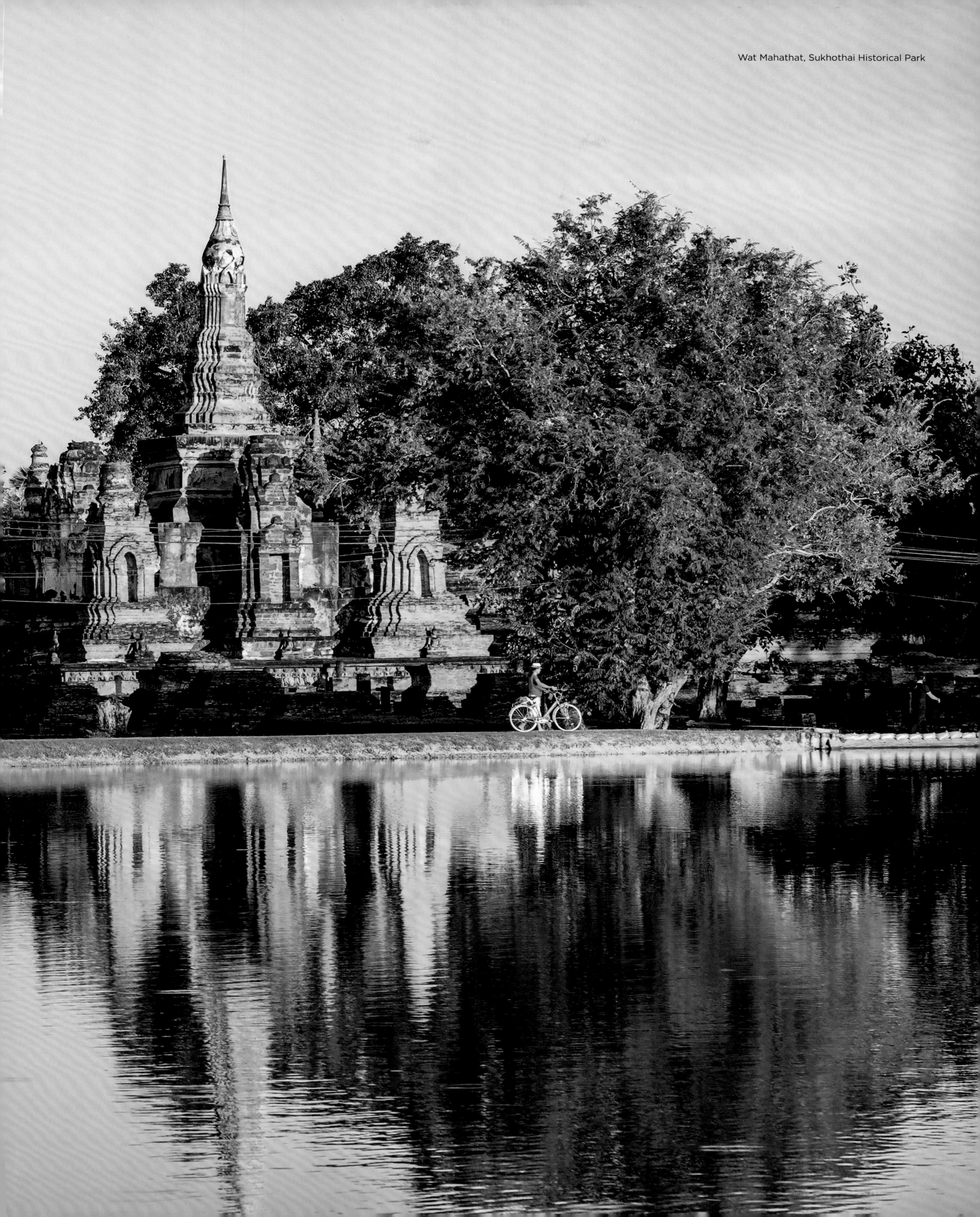

Kamphaeng Phet, Phitsanulok & Phetchabun

Khlong Lan National Park, Kamphaeng Phet

Wat Phra Kaew, Kamphaeng Phet Historical Park

Kamphaeng Phet, Phitsanulok & Phetchabun

These provinces, located in the centre of Thailand, captivate with their most diverse facets and nuances: populated mountainous regions, atmospheric stretches of water, centuries-old Buddha figures and panoramic views that enchant. Thung Salaeng Luang, one of the largest national parks in the region, has an area of more than 1200 km² and spreads across the provinces of Phitsanulok and Phetchabun.

Kamphaeng Phet, Phitsanulok & Phetchabun

Ces provinces situées au centre de la Thaïlande séduisent par leur diversité. On y est enchanté par les régions montagneuses peuplées, les étendues d'eau féériques, les bouddhas centenaires et les vues panoramiques. L'un des plus grands parcs nationaux de la région, Thung Salaeng Luang, a une superficie de plus de 1200 km² et s'étend sur les provinces de Phitsanulok et Phetchabun.

Kamphaeng Phet, Phitsanulok & Phetchabun

Die im Zentrum Thailands ansässigen Provinzen bestechen durch unterschiedlichste Facetten und Nuancen. Besiedelte Bergregionen, stimmungsvolle Gewässer, jahrhundertealte Buddhafiguren und Panoramaaussichten, die verzaubern. Einer der größten Nationalparks der Region heißt Thung Salaeng Luang, hat eine Fläche von mehr als 1200 km² und breitet sich über die Provinzen Phitsanulok und Phetchabun aus.

Kamphaeng Phet, Phitsanulok & Phetchabun

Las provincias situadas en el centro de Tailandia cautivan con las más diversas facetas y matices. Regiones montañosas pobladas, aguas atmosféricas, figuras de Buda centenarias y vistas panorámicas que encantan. Thung Salaeng Luang es uno de los parques nacionales más grandes de la región , tiene una superficie de más de 1200 km² y se extiende a través de las provincias de Phitsanulok y Phetchabun.

Kamphaeng Phet, Phitsanulok & Phetchabun

Le province situate nel centro della Thailandia sono un luogo dalle mille sfaccettature: regioni montuose, sede di numerosi villaggi, corsi d'acqua suggestivi, secolari statue del Buddha e panorami incantevoli. Uno dei più grandi parchi nazionali della regione è quello di Thung Salaeng Luang, che con una superficie di oltre 1.200 km² si estende tra le province di Phitsanulok e Phetchabun.

Kamphaeng Phet, Phitsanulok & Phetchabun

De provincies in het hart van Thailand fascineren door zeer uiteenlopende aspecten en nuances. Bevolkte berggebieden, sfeervolle wateren, eeuwenoude Boeddhabeelden en betoverende panoramische uitzichten. Een van de grootste nationale parken in de regio is Thung Salaeng Luang, heeft een oppervlakte van ruim 1200 km² en spreidt zich uit over de provincies Phitsanulok en Phetchabun.

Wat Phra Kaew, Kamphaeng Phet Historical Park

Kamphaeng Phet History Park

Declared a UNESCO World Heritage Site in 1991, the park has dozens of temples dating from the 13th to 15th centuries and stretches east of the Ping River. The mighty ruins with sitting Buddha statues, stone animal figures and bell-shaped chedis are relics from a glorious time when Kamphaeng Phet was one of the three largest cities of the kingdom of Sukhothai.

Parc historique de Kamphaeng Phet

Le parc, inscrit au patrimoine mondial de l'UNESCO en 1991, compte des dizaines de temples datant du XIIIᵉ au XVᵉ siècle et s'étend en grande partie à l'est de la rivière Ping. Les ruines imposantes, dont des statues de Bouddha assis, des animaux en pierre et des chedis en forme de clochettes sont les reliques d'une époque glorieuse où Kamphaeng Phet était l'une des trois plus grandes villes du royaume de Sukhothai.

Geschichtspark Kamphaeng Phet

Das von der UNESCO 1991 zum Weltkulturerbe erklärte Parkareal mit dutzenden Tempeln aus dem 13. bis 15. Jahrhundert erstreckt sich weitgehend östlich des Ping-Flusses. Die mächtigen Ruinen mit sitzenden Buddhastatuen, steinernen Tierfiguren und glockenförmigen Chedis sind Relikte aus einer vergangenen ruhmreichen Zeit, als Kamphaeng Phet zu den drei größten Städten des Sukhothai-Königreiches zählte.

Kamphaeng Phet Historical Park

Parque Histórico de Kamphaeng Phet

La zona del parque, que fue declarada Patrimonio de la Humanidad por la UNESCO en 1991, cuenta con decenas de templos que datan de los siglos XIII al XV y se extiende en gran parte al este del río Ping. Las poderosas ruinas con estatuas de Buda, figuras de animales de piedra y chedis en forma de campana son reliquias de una época gloriosa cuando Kamphaeng Phet era una de las tres ciudades más grandes del reino Sukhothai.

Parco sStorico Kamphaeng Phet

L'area del parco, dichiarata patrimonio dell'umanità dall'UNESCO nel 1991, comprende decine di templi edificati tra il XIII e il XV secolo e si estende in gran parte a est del fiume Ping. Le imponenti rovine, con le statue raffiguranti il Buddha seduto, figure di animali di pietra e chedi a forma di campana, sono resti di un passato glorioso, quando Kamphaeng Phet era una delle tre più grandi città del regno Sukhothai.

Historisch park Phet Kamphaeng

Het park, dat de Unesco in 1991 tot werelderfgoed uitriep, telt tientallen tempels uit de 13e tot 15e eeuw en strekt zich grotendeels ten oosten van de rivier de Ping uit. De machtige ruïnes met zittende Boeddhabeelden, stenen dierenfiguren en klokvormige chedi's zijn overblijfselen uit lang vervlogen tijden, toen Kamphaeng Phet een van de drie grootste steden van het koninkrijk Sukhothai was.

Phu Hin Rong Kla National Park, Phitsanulok

Wat Phra Sri Rattana Mahathat, Phitsanoluk

Pink-Lipped Habenaria, Phu Hin Rong Kla National Park, Phitsanulok

Phu Hin Rong Kla National Park

Diverse flora and fauna in all shapes and colours characterize this approximately 300 km² national park in Phitsanulok province. The highest point is the 1820 m high Phu Man Khao. Between lush green ferns and moss-covered rocks, picturesque waterfalls such as Man Dang find their way through the tropical landscape. Hiking trails run through the park, which was opened in 1984, and offer visitors a wonderful opportunity to experience magnificent nature at close range.

Parc national de Phu Hin Rong Kla

Ce parc national d'environ 300 km², dans la province de Phitsanulok, présente une flore et une faune très diverses. Son point culminant est le Phu Man Khao, haut de 1820 m. Des chutes d'eau pittoresques comme le Man Dang se frayent un chemin à travers un paysage tropical, entre les plants de fougères vertes luxuriantes et les rochers recouverts de mousse. Des sentiers de randonnées pédestres traversent le parc, ouvert en 1984, et offrent aux visiteurs une merveilleuse occasion de faire l'expérience de cette nature magnifique.

Phu Hin Rong Kla Nationalpark

Mannigfaltige Flora und Fauna in allen Formen und Farben zeichnen den rund 300 km² großen Nationalpark in der Provinz Phitsanulok aus. Der höchste Erhebung stellt der 1820 m hohe Phu Man Khao dar. Zwischen sattgrünen Farngewächsen und mit Moos bewachsenen Felsen finden immer wieder pittoreske Wasserfälle, wie der Man Dang, ihren Weg durch die tropische Landschaft. Wanderwege durchziehen den 1984 eröffneten Park und bieten Besuchern eine herrliche Möglichkeit die grandiose Natur hautnah erleben zu können.

Man Dang Waterfall, Phu Hin Rong Kla National Park, Phitsanulok

Parque nacional Phu Hin Rong Kla

Diversas especies de flora y fauna de todas las formas y colores caracterizan los aproximadamente 300 km² de este gran parque nacional de la provincia de Phitsanulok. El punto más alto es el Phu Man Khao de 1820 m de altura. Entre exuberantes helechos verdes y rocas cubiertas de musgo, pintorescas cascadas como la de Man Dang siguen encontrando su camino a través del paisaje tropical. El parque, inaugurado en 1984, cuenta con senderos que ofrecen a los visitantes una maravillosa oportunidad de experimentar de cerca la magnífica naturaleza.

Parco nazionale Phu Hin Rong Kla

Il grande parco nazionale nella provincia di Phitsanulok, con i suoi 300 km² di superficie, è caratterizzato da una flora e una fauna con forme e colori tra i più vari. Il monte Phu Man Khao, con i suoi 1.820 m, è il punto più alto del parco. Tra piante di felce e rocce ricoperte di muschio, alcune pittoresche cascate, come quella di Man Dang, caratterizzano il paesaggio tropicale. I sentieri escursionistici che attraversano il parco, inaugurato nel 1984, offrono ai visitatori una splendida opportunità di vivere da vicino la magnifica natura.

Nationaal park Phu Hin Rong Kla

Een afwisselende flora en fauna in alle vormen en kleuren kenmerken het ongeveer 300 km² grote nationale park in de provincie Phitsanulok. Het hoogste punt is de 1820 meter hoge Phu Man Khao. Tussen weelderige groene varens en met mos bedekte rotsen blijven pittoreske watervallen zoals de Man Dang hun weg vinden door het tropische landschap. Wandelpaden lopen door het in 1984 geopende park en bieden bezoekers een schitterende gelegenheid om de grandioze natuur van dichtbij te ervaren.

Phu Hin Rong Kla National Park, Phitsanulok

Koakho and Phu Tub Mountain, Phetchabun

Wat Phra Sorn Kaew, Khao Kho, Phetchabun

Thung Salaeng Luang National Park, Khao Kho, Phetchabun

Thung Salaeng Luang National Park

Tropical deciduous forests, lowland scrub and extensive meadows dominate this national park in the province of Phitsanulok. Depending on the time of day, nature presents itself with its most beautiful and colourful side. The hilly landscape with its outstanding individual mountains reaches heights of up to 1000 m. The reserve is an ideal area for trekking and bird watching. More than 190 species of birds are officially known, including eagles, owls, and hornbills, as well as the Siamese fireback pheasant, which is the national bird of Thailand. Larger wild animals such as elephants and deer can be seen at salt licks and water holes.

Parc national de Thung Salaeng Luang

Ce parc national de la province de Phitsanulok comprend des forêts tropicales de feuillus, des broussailles et de vastes prairies. À certaines heures du jour, la nature s'y présente sous son aspect le plus beau et coloré. Le paysage vallonné, avec des montagnes exceptionnelles, atteint des hauteurs allant jusqu'à 1000 m. La réserve est un endroit idéal pour le trekking et l'observation des oiseaux. Plus de 190 espèces y ont été officiellement recensées ; on y trouve ainsi des aigles, des hiboux, des calaos et des faisans siamois, considérés comme les oiseaux nationaux de Thaïlande. De plus grands animaux sauvages, tels que des éléphants ou des cerfs, peuvent être aperçus dans les salines et les bassins.

Thung Salaeng Luang Nationalpark

Tropische Laubwälder, Tiefland-Gestrüpp und ausgedehnte Wiesen bestimmen den Nationalpark in der Provinz Phitsanulok. Je nach Tageszeit präsentiert sich die vielfältige Natur von ihrer schönsten und farbenprächtigsten Seite. Die hügelige Landschaft mit einzelnen herausragenden Bergen erreicht Höhen bis rund 1000 m. Das Schutzgebiet ist ein idealer Ort zum Trekking und zur Vogelbeobachtung. Weit über 190 Vogelarten sind offiziell bekannt, darunter Adler, Eulen, der Nashornvogel oder der Siamesische Feuerrückenfasan, der als Nationalvogel Thailands gilt. Größere Wildtiere wie Elefanten und Hirsche trifft man an Salzlecken und Wasserlöchern.

Thung Salaeng Luang National Park, Khao Kho, Phetchabun

Parque nacional Thung Salaeng Luang

Bosques caducifolios tropicales, matorrales de tierras bajas y extensas praderas dominan el parque nacional en la provincia de Phitsanulok. Dependiendo de la hora del día, la naturaleza se presenta desde su lado más bello y colorido. El paisaje montañoso tiene montañas excepcionales, algunas de las cualesalcanzan alturas de hasta 1000 m. La reserva es un lugar ideal para el trekking y la observación de aves. Se conocen oficialmente más de 190 especies de aves, incluyendo águilas, búhos, el cálao y el faisán siamés, considerado el ave nacional de Tailandia. Animales silvestres más grandes como elefantes y ciervos se pueden encontrar en salinas y pozos de agua.

Parco nazionale di Salaeng Luang

Foreste tropicali, arbusti lungo le pianure e vasti prati dominano questo parco nazionale nella provincia di Phitsanulok. Durante l'arco della giornata, al variare della luce, la natura mostra tutte le sfaccettature dei suoi colori. Il paesaggio collinare, con alcuni picchi isolati, raggiunge altezze di 1.000 m. La riserva è un luogo ideale per il trekking e il birdwatching. Qui vivono ben oltre 190 specie di uccelli, tra cui l'aquila, la civetta, il bucero bicorne e il fagiano siamese, considerato l'uccello simbolo della Thailandia. Nei pressi di pietre saline e fosse d'acqua si incontrano spesso animali selvatici di grandi dimensioni, come elefanti e cervi.

Nationaal park Thung Salaeng Luang

Tropische loofbossen, struiken in het laagland en uitgestrekte weiden kenmerken het nationale park in de provincie Phitsanulok. Afhankelijk van het moment van de dag toont de natuur zich van haar mooiste en kleurrijkste kant. Het heuvelachtige landschap met her en der een hoog oprijzende berg bereikt hoogten tot ongeveer 1000 meter. Het natuurreservaat is een ideale plek om te wandelen en vogels te observeren. Er zijn meer dan 190 vogelsoorten officieel bekend, waaronder arenden, uilen, de neushoornvogel en de gekuifde vuurrugfazant, die wordt beschouwd als de nationale vogel van Thailand. Grotere wilde dieren zoals olifanten en herten zijn te vinden bij likplaasten en waterpoelen.

Loei & Nong Khai

Mekong River, Chiang Khan, Loei

Mekong River, Nong Khai

Loei & Nong Khai

The Mekong is the lifeline of both of these northern provinces as it meanders along the border between Thailand and Laos. Picturesque mountains such as Doi Phuwae in Loei rise up in this region and charming river landscapes display their beauty. Nong Khai is a culturally interesting city with holy sculptures and its colonial era French quarter.

Loei & Nong Khai

Le Mékong est la ligne de vie des deux provinces du nord, et serpente le long de la frontière entre la Thaïlande et le Laos. Des montagnes pittoresques comme le Doi Phuwae, à Loei, se dressent dans cette région et les paysages fluviaux pittoresques y sont de toute beauté. Nong Khai est une ville culturellement intéressante, notamment pour ses sculptures sacrées et son quartier français, autrefois colonial.

Loei & Nong Khai

Der Mekong ist die Lebensader beider Nordprovinzen. Er schlängelt sich entlang der Grenze zwischen Thailand und Laos. In dieser Region erheben sich pittoreske Gebirge wie der Doi Phuwae in Loei und malerische Flusslandschaften zeigen ihre ganze Schönheit. Eine kulturell interessante Stadt ist Nong Khai mit heiligen Skulpturen und dem einst durch die Kolonialisten geprägten französischen Viertel.

Phu Pa Por Hill, Loei

Loei & Nong Khai

El Mekong es la arteria de ambas provincias del norte.
Serpentea a lo largo de la frontera entre Tailandia y
Laos. Montañas pintorescas como la Doi Phuwae, en
Loei, se erigen en esta región y pintorescos paisajes
fluviales muestran toda su belleza. Nong Khai es
una ciudad culturalmente interesante,con esculturas
sagradas y su antiguo barrio colonial francés.

Loei & Nong Khai

Il Mekong è un'arteria importante per entrambe
le province settentrionali. Il fiume si snoda lungo
il confine con il Laos. Monti pittoreschi come il
Doi Phuwae a Loei si ergono in questa regione e i
paesaggi fluviali mostrano tutta la loro bellezza. Una
città di notevole interesse è Nong Khai, con le sue
sculture sacre e il vecchio quartiere coloniale francese.

Loei & Nong Khai

De Mekong is de levensader van beide noordelijke
provincies. Hij slingert langs de grens tussen Thailand
en Laos. In deze regio verrijzen pittoreske gebergten
zoals de Doi Phuwae in Loei en laten schilderachtige
rivierlandschappen al hun schoonheid zien. Een
cultureel interessante stad is Nong Khai, met heilige
beelden en de ooit door de kolonisatoren beïnvloede
Franse wijk.

Phu Kradueng Natonal Park, Loei

Cherry Blossom Flower, Phu Lom Lo, Loei

Mekong River, Nong Khai

Traditional fishing

Many Thais still feed their families with fishing. Throughout the country, people use ancient knowledge passed down from generation to generation to land handsome specimens of Siamese carp or catfish, using simple but effective methods. Huge lakes and the coastal region provide hope for a rich haul.

Pêche traditionnelle

De nombreux Thaïlandais nourrissent encore leurs familles grâce à la pêche. Partout dans le pays, les gens utilisent des connaissances anciennes transmises de génération en génération pour débusquer des spécimens de carpes siamoises ou de poissons-chats, en utilisant des méthodes simples mais efficaces. Les grands lacs et la région côtière offrent de riches et larges opportunités de pêche.

Traditioneller Fischfang

Viele Thailänder ernähren auch heute noch ihre Familien mit dem Fischfang. Im ganzen Land nutzen die Menschen uraltes, von Generation zu Genration weiter vermitteltes Wissen, um mit einfachen, aber effektiven Methoden stattliche Exemplare eines Siam-Karpfen oder eines Wels an Land zu ziehen. Riesige Seen und die Küstenregion lassen auf einen reiche Beute hoffen.

Pesca tradicional

Muchos tailandeses siguen alimentando a sus familias con la pesca. En todo el país, la gente utiliza los antiguos conocimientos transmitidos de generación en generación para obtener hermosos ejemplares de carpas o bagres siameses utilizando métodos sencillos pero eficaces. Enormes lagos, además de la región costera alimentan la esperanza de capturar una deliciosa presa.

Pesca tradizionale

Molti thailandesi vivono ancora di pesca. In tutto il paese, la gente utilizza antiche conoscenze tramandate di generazione in generazione per catturare con metodi semplici ma efficaci carpe o pesci gatto siamese. I grandi laghi e la regione costiera promettono ricchi bottini.

Traditionele visserij

Veel Thai voorzien nog in hun levensonderhoud met de visvangst. Overal in het land gebruiken mensen de oeroude kennis die van generatie op generatie is doorgegeven om prachtige Siamese reuzenkarpers of meervallen aan land te trekken met behulp van eenvoudige maar effectieve methoden. In de grote meren en de kuststreek wacht een rijke vangst.

Doi Phuwae Viewpoint, Loei

Sala Kaew Ku Sculpture Park, Nong Khai

Sala Kaew Ku Sculpture Park

Mystical stone beings from another world, the Hindu and Buddhist world of gods. Seven-headed snake statues, called naga, or 12-armed and multi-faced saints hide among lush tropical vegetation. Dozens of these sculptures, some of them more than 20 m high, are spread throughout the park on the banks of the Mekong River, about 5 km from Nong Khai. This unusual sight was created by the artist Luang Pu Bunluea Sulilat, who also created a similar sculpture park on the Laotian side of the river.

Parc des sculptures de Sala Kaew Ku

Des êtres mystiques en pierre issus des cultures hindou et bouddhiste, des statues de serpents à sept têtes, appelés naga, ou des saints à 12 bras et à plusieurs visages se cachent parmi une végétation tropicale luxuriante. Des dizaines de sculptures, dont certaines de plus de 20 m de haut, sont ainsi réparties dans ce parc sur les rives du Mékong, à environ 5 km de Nong Khai. Ce spectacle insolite a été créé par l'artiste Luang Pu Bunluea Sulilat, déjà à l'origine d'un parc de sculptures similaire au Laos.

Sala Kaew Ku Skulpturen-Park

Mystische steinerne Wesen aus seiner anderen Welt, aus einer hinduistisch-buddhistischen Götterwelt. Siebenköpfige Schlangenstatuen, Naga genannt, oder 12-armige und mehrgesichtige Heilige verstecken sich zwischen üppiger tropischer Vegetation. Dutzende dieser teils über 20 m hohen Skulpturen verteilen sich in dem etwa 5 km von Nong Khai entfernten, am Ufer des Mekong liegenden Parks. Geschaffen hat diese ungewöhnliche Sehenswürdigkeit der Künstler Luang Pu Bunluea Sulilat, der auch schon auf laotischer Seite einen ähnlichen Skulpturenpark ins Leben rief.

Sala Kaew Ku Sculpture Park, Nong Khai

Parque escultórico Sala Kaew Ku
Seres místicos de piedra que proceden de otro mundo, de uno de dioses hindúes y budistas. Estatuas de serpientes de siete cabezas, llamadas Naga, o santos de 12 brazos y múltiples caras se esconden entre la exuberante vegetación tropical. Decenas de estas esculturas, algunas de más de 20 m de altura, se distribuyen en el parque a orillas del río Mekong, a unos 5 km de Nong Khai. Este espectáculo inusual fue creado por el artista Luang Pu Bunluea Sulilat, que ya había creado un parque de esculturas similar en territoriolaosiano.

Parco delle sculture di Sala Kaew Ku
Figure mistiche in pietra come esseri provenienti da un altro mondo, un mondo di dèi indù e buddisti: statue di serpenti a sette teste, le cosiddette naga, o santi a dodici braccia e dai volti molteplici si nascondono tra la lussureggiante vegetazione tropicale. Decine di sculture di questo tipo, alcune delle quali alte più di 20 m, sono sparse nel parco sulle rive del fiume Mekong, a circa 5 km da Nong Khai. Questo spettacolo insolito è stato creato dall'artista Luang Pu Bunluea Sulilat, ideatore di un altro parco di sculture simile dall'altra parte del fiume, in Laos.

Beeldenpark Sala Kaew Ku
Mystieke stenen wezens uit een andere wereld, uit een hindoeïstisch-boeddhistische godenwereld. Zevenkoppige slangenbeelden, naga genaamd, of twaalfarmige en meerhoofdige heiligen staan verscholen in de weelderige tropische vegetatie. Tientallen van deze sculpturen, waarvan sommige meer dan 20 meter hoog zijn, staan verspreid in het park aan de oevers van de Mekong, op circa 5 km afstand van Nong Khai. Deze ongewone bezienswaardigheid is gecreëerd door de kunstenaar Luang Pu Bunluea Sulilat, die een vergelijkbaar beeldenpark aan de kant van Laos ontwierp.

ELEPHANTS

Elephants have held a very special position in Thailand for centuries. The majestic pachyderms are appreciated and revered, and are only very rarely used as work animals. In the national parks of the country some wild specimens still live today. The elephant temple Wat Chang Lom in Sukhothai displays specimens carved out of stone.

ÉLÉPHANTS

Les éléphants occupent une place très spéciale en Thaïlande, et ce depuis des siècles. Les majestueux pachydermes sont appréciés, vénérés et très rarement utilisés comme animaux de travail. Encore aujourd'hui, quelques spécimens sauvages vivent dans les parcs nationaux du pays. Le temple de l'éléphant Wat Chang Lom à Sukhothai en présente des spécimens sculptés dans la pierre.

ELEFANTEN

Elefanten besitzen in Thailand schon seit Jahrhunderten eine ganz besondere Stellung. Die majestätischen Dickhäute werden geschätzt, verehrt und nur noch sehr selten als Arbeitstiere eingesetzt. In den Nationalparks des Landes leben heute noch einige wilde Exemplare. Aus Stein gehauene Exemplare zeigt der Elefantentempel Wat Chang Lom in Sukhothai.

ELEFANTES

Los elefantes han ocupado una posición muy especial en Tailandia durante siglos. Los paquidermos majestuosos son apreciados, venerados y muy apenas utilizados como animales de trabajo. En los parques nacionales del país viven aún algunos ejemplares salvajes. El templo de elefantes de Wat Chang Lom, en Sukhothai, muestra especímenes tallados en piedra.

ELEFANTI

Già da secoli gli elefanti occupano una posizione molto speciale in Thailandia. I maestosi pachidermi sono apprezzati, venerati e solo molto raramente utilizzati come animali da lavoro. Nei parchi nazionali del paese vivono ancora oggi alcuni esemplari selvatici. Il Tempio dell'Elefante, Wat Chang Lom, a Sukhothai mostra esemplari scolpiti in pietra.

OLIFANTEN

Olifanten nemen al eeuwenlang een heel bijzondere positie in Thailand in. De majestueuze dikhuiden worden gewaardeerd, vereerd en nog maar heel weinig als werkdieren ingezet. In de nationale parken van het land leven nog enkele wilde dieren. De olifantentempel Wat Chang Lom in Sukhothai toont uit steen gehouwen exemplaren.

161

Udon Thani, Sakon Nakhon & Nakhon Phanom

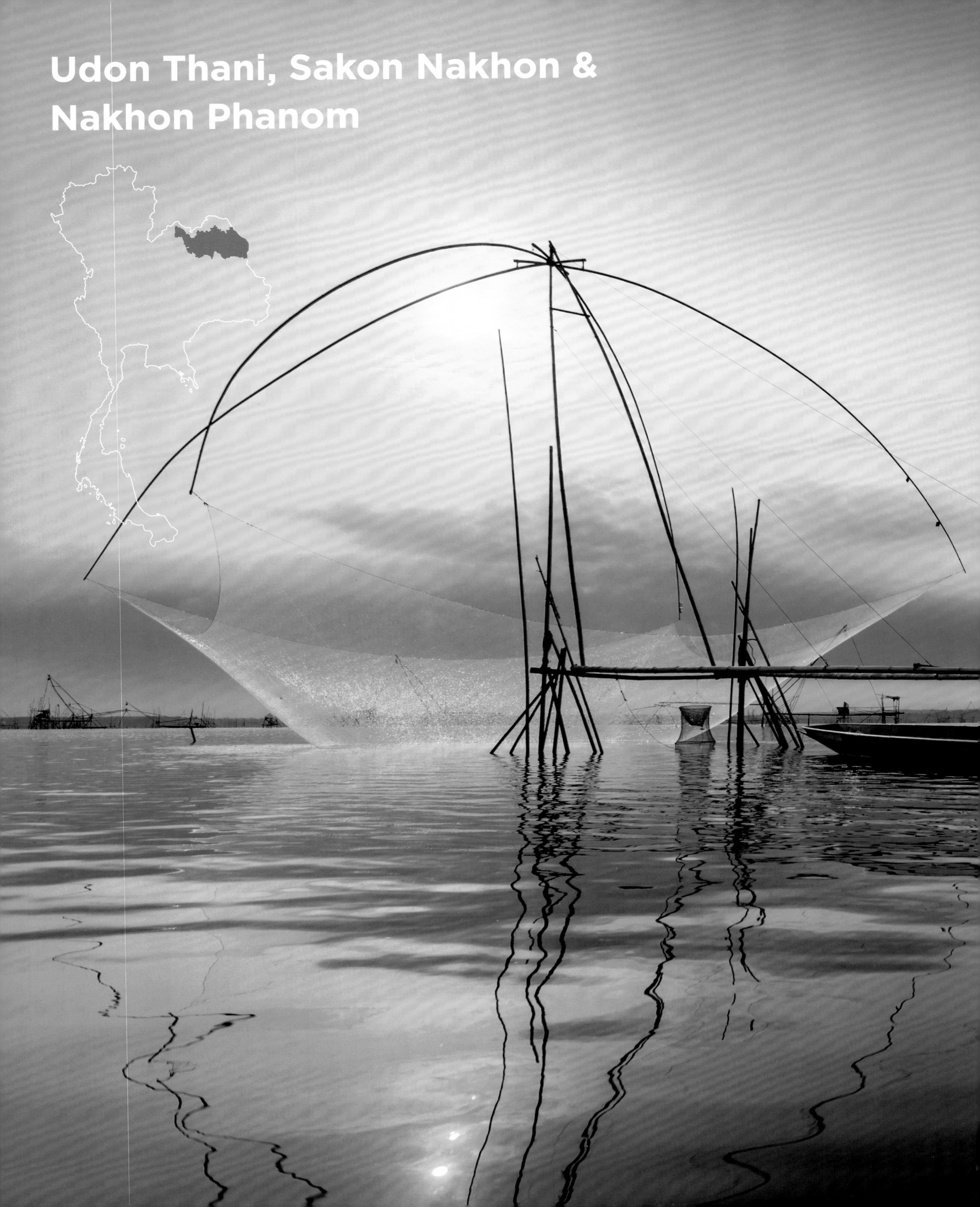

Huai Luang Reservoir, Udon Thani

Phu Phra Bat National Park, Udon Thani

Udon Thani, Sakon Nakhon & Nakhon Phanom

Phu Phra Bat Park in Udon Thani is known for its rock paintings and Wat Phra Phutthabat Bua Bo, which attracts many pilgrims who want to visit the footprint of Buddha. The Huai Luang Reservoir is more for activities such as fishing, canoeing and rafting. The local population benefits from its power generation and the abundance of fish. Nong Han Lake, measuring over 120 km², was built without any human help, which makes it Thailand's largest natural freshwater lake. The Third Thai-Laotian Friendship Bridge also has stately dimensions. This construction stretches over the Mekong and has a length of almost 1.5 km.

Udon Thani, Sakon Nakhon & Nakhon Phanom

Le parc historique de Phu Phra Bat, à Udon Thani, est célèbre pour ses peintures rupestres ainsi que pour le temple de Wat Phra Phutthabat Bua Bo, qui attire de nombreux pèlerins venus admirer l'empreinte de pas de Bouddha. Au réservoir de Huai Luang, on peut pratiquer des activités telles que la pêche, le canoë-kayak et le rafting. Pour la population locale, il est également synonyme de production d'électricité et de poissons en abondance. Le lac Nong Han, d'une superficie de plus de 120 km², n'est pas une construction humaine, ce qui en fait le plus grand lac d'eau douce naturel de Thaïlande. Le troisième pont de l'amitié, qui relie la Thaïlande et le Laos, a également des dimensions majestueuses : la structure, qui s'étend sur le Mékong, est longue de près de 1,5 km.

Udon Thani, Sakon Nakhon & Nakhon Phanom

Der Phu Phra Bat Park in Udon Thani ist bekannt für seine Felsmalereien und den Wat Phra Phutthabat Bua Bo, der viele Pilger anzieht, die dem Fußabdrucks Buddhas einen Besuch abstatten möchten. Das Huai Luang Reservoir steht eher für Aktivitäten wie Angeln, Kanufahren und Rafting. Die einheimische Bevölkerung profitiert von der Stromerzeugung und dem Fischreichtum. Ganz ohne menschliche Hilfe entstand der über 120 km² messende Nong Han. Damit ist er der größte natürliche Süßwassersee Thailands. Über ebenfalls stattliche Ausmaße verfügt die Dritte Thailändisch-Laotische Freundschaftsbrücke. Die sich über den Mekong spannende Konstruktion hat eine Länge von knapp 1,5 km.

Phu Phra Bat National Park, Udon Thani

Udon Thani, Sakon Nakhon & Nakhon Phanom

El Parque del Murciélago de Phu Phra en Udon Thani es conocido por sus pinturas rupestres y al templo Wat Phra Phutthabat Bua Bo, que atrae a muchos peregrinos que quieren visitar la huella de Buda. El embalse de Huai Luang permite actividades como la pesca, el piragüismo y el rafting. La población local se beneficia de la generación de energía y de la abundancia de peces. El lago Nong Han, de más de 120 km², surgió sin que mediara la mano del hombre. Esto lo convierte en el lago natural de agua dulce más grande de Tailandia. El Tercer Puente de la Amistad entre Tailandia y Laos también tiene dimensiones majestuosas. La construcción, que se extiende sobre el Mekong, tiene una longitud de casi 1,5 km.

Udon Thani, Sakon Nakhon & Nakhon Phanom

Il parco di Phu Phra Bat a Udon Thani è noto per le sue pitture rupestri e per il Wat Phra Phutthabat Bua Bo, che attira molti pellegrini in visita per l'impronta del piede del Buddha qui conservata. Presso il bacino artificiale di Huai Luang è possibile svolgere attività come la pesca, il canottaggio e il rafting. La popolazione locale beneficia della produzione di energia e dell'abbondanza di pesce. Il lago di Nong Han, con una superficie di oltre 120 km², non è invece opera umana ma è il più grande lago naturale d'acqua dolce della Thailandia. Anche il terzo ponte dell'amicizia thailandese-laotiana ha dimensioni maestose: la costruzione sul Mekong ha una lunghezza di quasi 1,5 km.

Udon Thani, Sakon Nakhon & Nakhon Phanom

Het Phu Phra Bat Park in Udon Thani staat bekend om zijn rotstekeningen en de Wat Phra Phutthabat Bua Bo, waar veel pelgrims op afkomen die de voetafdruk van Boeddha willen bezoeken. Het Huai Luang Reservoir staat eerder voor activiteiten als vissen, kanoën en raften. De lokale bevolking profiteert van de elektriciteitsopwekking en de overvloed aan vis. Het Nong Han-meer, met een oppervlakte van meer dan 120 km², is ontstaan zonder menselijke hulp. Daarmee is het Thailands grootste natuurlijke zoetwatermeer. Ook de Derde Thais-Laotiaanse Vriendschapsbrug heeft statige afmetingen. De constructie, die de Mekong overspant, heeft een lengte van bijna 1,5 km.

Red Lotus, Nong Han Lake, Sakon Nakhon

Third Thai Lao Friendship Bridge, Nakhon Phanom

Wat Phra That Choeng Chum, Sakon Nakhon

Charming Temples

The former Khmer city of Sakon Nakhon boasts a brilliant attraction from days long gone: the temple complex Wat Phra That Choeng Chum. The buildings are richly decorated and embellished with precious building materials and portraits of the Buddha, which are presented in the Khmer and Lao styles. Wat Phra That Panom in the northeast of Thailand is hardly less splendid. The faithful worship it as one of the most important of its kind in this remote region on the border with Laos. The centrepiece is the over 50 m high chedi, decorated with precious stones and gold, surrounded by dozens of small Buddha figures.

Temples de charme

L'ancienne ville khmère de Sakon Nakhon a connu une époque fastueuse, ainsi que le prouve le complexe du temple Wat Phra That Choeng Chum. Les bâtiments y sont richement agrémentés, conçus avec des matériaux précieux et décorés de représentations du Bouddha, présentés dans les styles khmer et laotien. Wat Phra That Panom, au nord-est de la Thaïlande, est à peine moins splendide. Pour les fidèles, ce temple est l'un des plus importants dans cette région éloignée frontalière du Laos. La pièce maîtresse est le chedi de plus de 50 m de haut, décoré de pierres précieuses et d'or, entouré de dizaines de petites statues de Bouddha.

Bezaubernde Tempel

Die einstige Khmerstadt Sakon Nakhon verfügt über eine brillante Sehenswürdigkeit aus längst vergangenen Tagen. Es handelt sich dabei um den Tempelkomplex Wat Phra That Choeng Chum. Es zeigen sich reichverzierte und mit kostbaren Baumaterialen veredelte Gebäude und Buddhabildnisse, die im Khmer- und laotischen Stil angelegt sind. Kaum weniger prachtvoll erscheint der Wat Phra That Panom im Nordosten Thailands. Die Gläubigen verehren ihn als einen der wichtigsten seiner Art in dieser entlegenen Region an der Grenze zu Laos. Mittelpunkt ist der über 50 m hohe, mit Edelsteinen und Gold dekorierte Chedi, den dutzende kleine Buddhafiguren umringen.

Wat Phra That Panom, Nakhon Phanom

Templos con Encanto

La antigua ciudad jemer de Sakon Nakhon cuenta con un monumento estelar desde hace mucho tiempo. Es el complejo del templo Wat Phra That Choeng Chum. Los edificios están ricamente decorados y refinados con preciosos materiales de construcción y retratos de Buda, que están dispuestos en los estilos jemer y laosiano. El templo Wat Phra That Panom, en el noreste de Tailandia, no desmerece en absoluto. Los fieles lo adoran como uno de los más importantes de su clase en esta remota región fronteriza con Laos. La pieza central es el Chedi de más de 50 m de altura, decorado con piedras preciosas y oro, rodeado de docenas de pequeñas figuras de Buda.

Templi magici

La città di Sakon Nakhon, un tempo parte del regno Khmer, è la sede di un luogo sacro incredibile. Si tratta del complesso del tempio di Wat Phra That Choeng Chum. Gli edifici sono riccamente decorati con preziosi materiali, i ritratti del Buddha che qui si trovano sono stati realizzati in stile in parte khmer e in parte laotiano. Non meno maestoso il tempio di Wat Phra That Panom, nel nord-est della Thailandia. I fedeli lo venerano come uno dei più importanti del suo genere in questa remota regione al confine con il Laos. Il punto più importante è il chedi, alto oltre 50 m, decorato con pietre preziose e oro e circondato da decine di piccole statue del Buddha.

Betoverende tempels

De voormalige Khmerstad Sakon Nakhon beschikt over een briljante bezienswaardigheid uit lang vervlogen tijden: het tempelcomplex Wat Phra That Choeng Chum. Er staan rijkversierde, met kostbare bouwmaterialen opgesmukte gebouwen en Boeddhabeelden in de stijl van Laos en de Khmer. Wat Phra That Panom in het noordoosten van Thailand is niet minder prachtig. De gelovigen aanbidden het heiligdom als een van de belangrijkste van zijn soort in deze afgelegen regio op de grens met Laos. Het middelpunt is de meer dan 50 meter hoge chedi, die met edelstenen en goud is versierd en wordt omringd door tientallen kleine Boeddhabeelden.

Ubon Ratchathani

Sam Phan Bok

Ubon Ratchathani

This beautiful province spreads itself elegantly out along the so-called "Golden Triangle", on the border with Laos and Cambodia. One of the most scenic and extraordinary gems of the region is Sam Phan Bok, with its round polished rocks and thousands upon thousands of holes created over the years by the rushing waters of the Mekong River.

Ubon Ratchathani

Cette belle province s'étend élégamment le long du « Triangle d'or », à la frontière avec le Laos et le Cambodge. L'un des lieux les plus pittoresques et extraordinaires de la région s'appelle Sam Phan Bok. C'est une formation rocheuse étonnante : sur les bords de la rivière, le sol est parsemé de milliers de trous, créés au fil des ans par les eaux rugissantes de la rivière Mae.

Ubon Ratchathani

Diese bildschöne Provinz breitet sich elegant am sogenannten „Goldenen Dreieck" aus, an der Grenze zu Laos und Kambodscha. Eine der landschaftlich reizvollsten und außergewöhnlichsten Glanzstücke der Region trägt den Namen Sam Phan Bok. All die rundgeschliffenen Felsen und abertausenden Löcher erstanden durch das Jahr ein und Jahr aus dahinrauschende Wasser des Mekong.

Sam Phan Bok

Ubon Ratchathani

Esta hermosa provincia se extiende con elegancia
a lo largo del llamado "Triángulo de Oro", en la
frontera con Laos y Camboya. Uno de los lugares más
pintorescos y extraordinarios de la región se llama
Sam Phan Bok. Todas las rocas redondeadas por la
erosión y los miles y miles de orificios son el efecto de
las aguas murmurantes del río Mekong durante años.

Ubon Ratchathani

Questa provincia affascinante si estende
elegantemente lungo il cosiddetto "triangolo d'oro",
al confine con Laos e Cambogia. Una delle attrazioni
più interessanti e straordinarie della regione è il Sam
Phan Bok: le rocce lisce e tondeggianti e le migliaia
di buche sono state create dalle acque scroscianti del
fiume Mekong.

Ubon Ratchathani

Deze beeldschone provincie strekt zich elegant uit
langs de zogenaamde "Gouden Driehoek" op de
grens met Laos en Cambodja. Een van de mooiste
en qua landschap meest bijzondere hoogtepunten
van de regio draagt de naam Sam Phan Bok. Alle
rondgeslepen rotsen en de vele duizenden gaten erin
zijn ontstaan doordat door het water van de rivier de
Mekong er jaar na jaar overheen raasde.

Sam Phan Bok

Pre-historic rock paintings, Pha Taem National Park

Rock paintings and deep gorges

Thousands of years of flowing water has created this unique natural spectacle. The bizarrely shaped canyons of Sam Phan Bok are also known as "The Grand Canyon of Thailand". Similarly impressive are the 3000 to 4000 year old rock paintings in Pha Taem National Park. You can recognize human figures and different animal species such as elephants and fish.

Pinturas rupestres y gargantas profundas

Miles de años de agua en circulación crearon este espectáculo natural único. Los extraños cañones de Sam Phan Bok son también conocidos como "El Gran Cañón de Tailandia". Igualmente impresionantes son las pinturas rupestres de 3.000 a 4.000 años de antigüedad en el Parque nacional de Pha Taem. Se reconocen figuras humanas y diferentes especies animales como elefantes y peces.

Peintures rupestres et profondes gorges

Spectacle naturel unique forgé par des milliers d'années de courants, les étranges canyons de Sam Phan Bok sont également connus sous le nom de « Grand Canyon de Thaïlande ». Les peintures rupestres datant de 3 000 à 4 000 ans, dans le parc national de Pha Taem, sont tout aussi impressionnantes. On reconnaît des figures humaines et différentes espèces animales comme des éléphants et des poissons.

Pitture rupestri e gole profonde

Migliaia di anni di acqua che scorre hanno creato questo spettacolo naturale unico nel suo genere. I bizzarri canyon di Sam Phan Bok sono conosciuti anche come "il Grand Canyon della Thailandia". Altrettanto spettacolari sono le pitture rupestri risalenti a circa 3.000, 4.000 anni fa nel Parco nazionale di Pha Taem, che ritraggono figure umane e diverse specie animali tra cui elefanti e pesci.

Felsmalereien und tiefe Schluchten

Tausende Jahre dahinfließendes Wasser schuf dieses einzigartige Naturschauspiel. Die bizarr geformten Felsschluchten von Sam Phan Bok sind auch unter der Bezeichnung „Der Grand Canyon von Thailand" berühmt. Ähnlich eindrucksvoll behaupten sich die 3000 bis 4000 Jahre alten Felszeichnungen im Pha Taem Nationalpark. Man erkennt menschliche Figuren und verschiedene Tierarten wie Elefanten und Fische.

Rotsschilderingen en diepe kloven

Door het water dat hier al duizenden jaren stroomt, is dit unieke natuurspektakel ontstaan. De bizar gevormde ravijnen van Sam Phan Bok staan ook wel bekend als "de Grand Canyon van Thailand". Even indrukwekkend zijn de 3000 tot 4000 jaar oude rotstekeningen in het nationale park Pha Taem. Je kunt er menselijke figuren en verschillende diersoorten zoals olifanten en vissen in herkennen.

Wat Tham Heo Sin Chai

Nakhon Ratchasima, Buriram &
Nakhon Nayok

Makha Bucha Buddhist Memorial Park, Nakhon Nayok

Khao Yai National Park, Nakhon Ratchasima

Khao Yai National Park, Nakhon Ratchasima

Nakhon Ratchasima, Buriram & Nakhon Nayok
The three provinces extend over the Khorat Plateau, a hilly area used mainly for agriculture. The largest city is Nakhon Ratchasima, or Khorat for short, with about 140,000 inhabitants. Away from the cities there is the breathtaking starry sky in Khao Yai National Park, or historically interesting temples such as Prasat Mueang Tam.

Nakhon Ratchasima, Buriram & Nakhon Nayok
Les trois provinces s'étendent dans la zone du plateau vallonné de Khorat, principalement utilisé pour l'agriculture. La plus grande ville est Nakhon Ratchasima, qui en portait autrefois le nom, et qui compte environ 140 000 habitants. En dehors des villes, il est possible d'y admirer des ciels étoilés époustouflants dans le parc national de Khao Yai ou d'y visiter des temples historiquement intéressants comme Prasat Mueang Tam.

Nakhon Ratchasima, Buriram & Nakhon Nayok
Die drei Provinzen erstrecken sich im Bereich des Khorat-Plateaus, einer hügeligen Hochebene, die hauptsächlich für die Landwirtschaft genutzt wird. Die größte Stadt ist Nakhon Ratchasima, kurz Khorat, mit etwa 140.000 Einwohnern. Abseits der Städte zeigen sich ein atemberaubender Sternenhimmel im Khao Yai Nationalpark oder geschichtlich interessante Tempelanlagen, wie der Prasat Mueang Tam.

Nakhon Ratchasima, Buriram & Nakhon Nayok
Las tres provincias se extienden en la zona de la meseta de Khorat, una meseta montañosa utilizada principalmente para la agricultura. La ciudad más grande es Nakhon Ratchasima, o Khorat para abreviar, con unos 140.000 habitantes. Lejos de las ciudades hay un impresionante cielo estrellado en el Parque nacional de Khao Yai o templos de interés histórico como Prasat Mueang Tam.

Nakhon Ratchasima, Buriram & Nakhon Nayok
Le tre province si estendono nella zona dell'altopiano di Khorat, una regione collinare dove viene praticata principalmente l'agricoltura. La città più grande è Nakhon Ratchasima, chiamata anche Khorat, con circa 140.000 abitanti. Lontano dalle città è possibile ammirare un cielo stellato mozzafiato nel Parco nazionale di Khao Yai così come pure templi ricchi di storia, come quello di Prasat Mueang Tam.

Nakhon Ratchasima, Buriram & Nakhon Nayok
De drie provincies strekken zich uit over het gebied van het Khoratplateau, een heuvelachtige hoogvlakte die voornamelijk voor landbouw wordt gebruikt. De grootste stad is Nakhon Ratchasima, of kortweg Khorat, met ongeveer 140.000 inwoners. Ver weg van de steden kunt u een adembenemende sterrenhemel bewonderen in het nationale park Khao Yai of historisch interessante tempelcomplexen zoals Prasat Mueang Tam bezoeken.

Haew Narok Waterfall, Khao Yai National Park, Nakhon Ratchasima

Prasat Mueang Tham, Buriram

Prasat Mueang Tham, Buriram

Khmer Temple

In the 11th century, the Khmers built the Prasat
Mueang Tam, which has a square shape and four
gates. It is located about 50 km south of the
provincial capital Buri Ram. The main entrance
faces east, as do most Hindu and Buddhist temples
in Thailand. The four L-shaped ponds, overgrown
with water lilies, were created as a symbol of the
"Goddess of Water", who is regarded as the patron
saint of the temple. The nearby Phanom Rung is
even more magnificent and larger. It is one of the
most impressive holy sites of the Khmer era, and is
enthroned on the extinct volcano Khao Phanom Rung.

Temple Khmer

Au XIe siècle, les Khmers ont bâti le Prasat Mueang
Tam, temple carré à quatre portes. Il est situé à
environ 50 km au sud de la capitale provinciale Buri
Ram. L'entrée principale est orientée vers l'est, comme
dans la plupart des temples hindous et bouddhistes
du pays. Les quatre étangs en forme de L, recouverts
de nénuphars, sont des symboles de la « déesse de
l'eau », considérée comme la patronne du temple. Le
Phanom Rung tout proche est plus grand. Trônant
sur le volcan éteint Khao Phanom Rung, c'est l'un des
sanctuaires les plus impressionnants de l'ère khmère.

Tempel aus der Khmer-Zeit

Im 11. Jahrhundert erbauten die Khmer den Prasat
Mueang Tam, der eine quadratische Form aufweist
und mit vier Toren ausgestattet ist. Er befindet sich
etwa 50 km südlich der Provinzhauptstadt Buri Ram.
Der Haupteingang ist gen Osten gerichtet, wie in den
meisten hinduistischen und buddhistischen Tempeln
in Thailand. Die vier L-förmigen, mit Seerosen
bewachsenen Teiche wurden als Symbol für die
„Göttin des Wassers" angelegt, die als Schutzpatronin
des Tempels gilt. Der nahegelegene Phanom Rung
stellt sich prachtvoller und größer dar. Er gehört zu
den beeindruckensten Heiligtümern aus der Khmer-
Zeit und thront auf dem erloschen Vulkan Khao
Phanom Rung.

Phanom Rung, Buriram

Templo de la época jemer

En el siglo XI, los jemeres construyeron el Prasat Mueang Tam, que tiene forma cuadrada y cuatro entradas. Se encuentra a unos 50 km al sur de la capital de provincia Buri Ram. La entrada principal está orientada hacia el este, como en la mayoría de los templos hindúes y budistas de Tailandia. Los cuatro estanques en forma de ele, cubiertos de nenúfares, fueron diseñados como símbolo de la "diosa del agua", considerada la patrona del templo. El templo de Phanom Rung está cerca y resulta aún más magnífico y más grande. Es uno de los santuarios más impresionantes de la era jemer y está entronizado en el extinto volcán Khao Phanom Rung.

Tempio khmer

Nell'XI secolo, i Khmer costruirono il Prasat Mueang Tam, a pianta rettangolare, a cui si accede da quattro portali. Si trova a circa 50 km a sud del capoluogo di provincia Buri Ram. L'ingresso principale è rivolto a est, come nella maggior parte dei templi indù e buddisti in Thailandia. Le quattro vasche a forma di L, piene di ninfee, sono state create come simbolo della "dea dell'acqua", a cui è dedicato il tempio. Il vicino Phanom Rung, più imponente sia nello sfarzo che nelle dimensioni, è uno dei santuari più impressionanti dell'epoca khmer e si erge sui pendii del vulcano spento Khao Phanom Rung.

Tempels uit de Khmertijd

In de 11e eeuw bouwden de Khmer het tempelcomplex Prasat Mueang Tam, dat een vierkante vorm heeft en over vier poorten beschikt. Het ligt ongeveer 50 km ten zuiden van de provinciehoofdstad Buri Ram. De hoofdingang is op het oosten gericht, net als bij de meeste hindoeïstische en boeddhistische tempels in Thailand. De vier L-vormige, met waterlelies begroeide vijvers werden aangelegd als symbool van de "godin van het water", die als de beschermheilige van de tempel wordt beschouwd. Het nabijgelegen heiligdom Phanom Rung is magnifieker en groter. Het is een van de indrukwekkendste complexen uit de Khmertijd en staat boven op de uitgedoofde vulkaan Khao Phanom Rung.

MONKS

Monks dressed in an orange robe are not uncommon on Thailand's streets. They have devoted themselves entirely to Buddha's teaching and largely renounce material wealth and property. Among their few permitted possessions is their alms bowl, with which they often walk early in the morning to receive rice and other food from the faithful.

MOINES

Les moines, vêtus d'une robe orange, ne sont pas rares dans les rues de Thaïlande. Ils se sont entièrement consacrés à l'enseignement de Bouddha et renoncent largement à la prospérité et à la propriété. Parmi les rares possessions qui leur sont permises se trouve le bol d'aumône avec lequel ils se déplacent souvent tôt le matin pour recevoir du riz et d'autres aliments des fidèles.

MÖNCHE

In ein orangefarbenes Gewand gehüllte Mönche sind keine Seltenheit auf Thailands Straßen. Sie haben sich ganz und gar der Lehre Buddhas verschrieben und verzichten weitestgehend auf Wohlstand und Eigentum. Zu ihren wenigen erlaubten Besitztümer gehört die Almosenschale mit der sie oft frühmorgens unterwegs sind, um von den Gläubigen Reis und andere Speisen zu erhalten.

FRAILES

Los monjes vestidos con una túnica naranja son comunes en las calles de Tailandia. Se han dedicado por completo a las enseñanzas de Buda y renuncian en gran medida a la prosperidad y a la propiedad. Entre sus pocas posesiones permitidas está el cuenco de limosnas con el que a menudo viajan temprano por la mañana para recibir arroz y otros alimentos de los fieles.

MONACI

Con la loro tonaca arancione, i monaci sono una apparizione tutt'altro che rara per le strade della Thailandia. Seguaci dell'insegnamento del Buddha, essi rinunciano quasi del tutto alla prosperità e alla proprietà. Tra i pochi possedimenti consentiti c'è la ciotola dell'elemosina, usata spesso la mattina per ricevere riso e altro cibo dai fedeli.

MONNIKEN

Monniken gekleed in een oranje gewaad zijn beslist geen zeldzaamheid in het straatbeeld van Thailand. Zij hebben zich volledig gewijd aan de leer van Boeddha en zien grotendeels af van welvaart en bezit. Een van hun weinige toegestane bezittingen is de bedelschaal waarmee ze vaak vroeg in de ochtend op weg zijn om rijst en ander voedsel van de gelovigen in ontvangst te nemen.

Lopburi & Ayutthaya

Ayutthaya Historical Park

Lopburi, Ayutthaya & Ang Thong

Ayutthaya, the former capital of the kingdom of Siam entrances with its magnificent variety of centuries-old holy shrines. Lopburi's sights, including the Wat Phra Sri Rattana Mahathat, which was probably built in the 11th century, are hardly less gracefully received by visitors. Among them is Wat Phra Sri Rattana Mahathat, which probably originated in the 11th century. In Ang Thong, Wat Muang, inaugurated by King Bhumibol in 1986, inspires with its 93 m high golden Buddha statue.

Lopburi, Ayutthaya & Ang Thong

Ayutthaya, la antigua capital del reino tailandés de Siam, cautiva por su espléndida variedad de santuarios centenarios. No menos agraciados son los monumentos de Lopburi, que dan la bienvenida a los visitantes, entre ellos, el temploWat Phra Sri Rattana Mahathat, probablemente erigido en el siglo XI. En Ang Thong, se sitúa el templo Wat Muang, inaugurado por el rey Bhumibol en 1986. Esta obra destacapor la estatua dorada de Buda, de 93 m de altura.

Lopburi, Ayutthaya & Ang Thong

Ayutthaya, l'ancienne capitale du royaume thaïlandais du Siam, ensorcèle par sa grandiose variété de sanctuaires séculaires. La province de Lopburi propose également aux visiteurs de nombreuses curiosités, parmi lesquelles le temple de Wat Phra Sri Rattana Mahathat, datant probablement du XIᵉ siècle. À Ang Thong, le Wat Muang et sa statue en or du Bouddha, haute de 93 m et inaugurée par le roi Bhumibol en 1986, est particulièrement enthousiasmante.

Lopburi, Ayutthaya & Ang Thong

Ayutthaya, un tempo capitale del regno thailandese del Siam, colpisce per la varietà dei santuari secolari. Non meno stupefacenti le bellezze che attendono i visitatori a Lopburi, tra cui il tempio di Wat Phra Sri Rattana Mahathat, fondato probabilmente nel XI secolo. Ad Ang Thong si trova il Wat Muang, inaugurato dal re Bhumibol nel 1986, con una impressionante statua d'oro del Buddha alta 93 m.

Lopburi, Ayutthaya & Ang Thong

Ayutthaya, die einstige Hauptstadt des thailändischen Königsreichs Siam betört durch eine grandiose Vielfalt Jahrhunderte alter Heiligtümer. Kaum weniger anmutig empfangen Lopburis Sehenswürdigkeiten die Besucher, darunter der wohl im 11. Jahrhundert entstandene Wat Phra Sri Rattana Mahathat. In Ang Thong begeistert der 1986 von König Bhumibol eingeweihe Wat Muang mit der 93 m hohen goldenen Buddhastatue.

Lopburi, Ayutthaya & Ang Thong

Ayutthaya, de voormalige hoofdstad van het Thaise koninkrijk Siam, betovert door een grandioze verscheidenheid aan eeuwenoude heiligdommen. Niet minder gracieus verwelkomen de bezienswaardigheden van Lopburi de bezoekers, waaronder het waarschijnlijk in de 11e eeuw gebouwde Wat Phra Sri Rattana Mahathat. In Ang Thong boeit het door koning Bhumibol in 1986 ingehuldigde complex Wat Muang met zijn 93 meter hoge gouden Boeddhabeeld.

Prang Sam Yot, Lopburi

Prang Sam Yot, Lopburi

Wat Phra Sri Rattana Mahathat, Lopburi

Religious Relics

Monkeys wherever you look. Prang Sam Yot is known, if not notorious, for its huge flock of monkeys. The temple itself displays Hindu style and building elements, and its Buddhist figures were later added. In the centre of Lopburi, Wat Phra Sri Rattana Mahathat is also connected to Buddhism. Many of its religious buildings and figures have deteriorated over the centuries.

Reliquias Religiosas

Monos por doquier. El Prang Sam Yot es conocido, si no notorio, por su enorme bandada de monos residentes. El templo en sí mismo muestra el estilo hindú y los elementos de construcción. Más tarde se añadieron las figuras budistas. El Wat Phra Sri Rattana Mahathat, en el centro de Lopburi, también pertenece al budismo. Muchos de sus edificios y figuras religiosas se deterioraron a lo largo de los siglos.

Reliques religieuses

Des singes à perte de vue : le temple de Prang Sam Yot est en effet bien connu pour abriter une immense communauté de macaques. Il est également un bon exemple de constructions de style hindou – les sculptures bouddhistes y ont été ajoutées par la suite. Le Wat Phra Sri Rattana Mahathat, au centre de Lopburi, est également un temple bouddhiste, mais beaucoup de ses édifices religieux et de ses statues se sont délabrés au fil des siècles.

Reliquie sacre

Scimmie ovunque si posi lo sguardo: il Prang Sam Yot è noto – o famigerato, se si preferisce – per l'elevato numero di scimmie che vi abitano. Il tempio mostra chiari segni dello stile indù anche negli elementi costruttivi. Le figure buddiste sono state aggiunte in seguito. Anche il Wat Phra Sri Rattana Mahathat, nel centro di Lopburi, è un tempio buddista. Molti dei suoi edifici e delle figure un tempo presenti sono andati in rovina nel corso dei secoli.

Religiöse Relikte

Affen, wohin man schaut. Der Prang Sam Yot ist bekannt, wenn nicht gar berüchtigt, für seine riesige dort lebende Affenschar. Der Tempel selbst zeigt hinduistische Stil- und Bauelemente. Die buddhistischen Figuren fügte man später hinzu. Dem Buddhismus zugehörig ist auch der Wat Phra Sri Rattana Mahathat im Zentrum von Lopburi. Viele seiner religiösen Gebäude und Figuren verfielen im Laufe der Jahrhunderte.

Religieuze overblijfselen

Apen, waar je ook kijkt. De Prang Sam Yot is bekend, zo niet berucht, om zijn enorme schare apen die daar leeft. De tempel zelf toont hindoeïstische stijl- en bouwelementen. De boeddhistische figuren werden later toegevoegd. Het tempelcomplex Wat Phra Sri Rattana Mahathat in het centrum van Lopburi is ook boeddhistisch. Veel religieuze gebouwen en beelden zijn in de loop der eeuwen in verval geraakt.

Bang Pa, Ayutthaya

Wat Phra Si Sanphet, Ayutthaya Historical Park

Wat Chai Wattanaram, Ayutthaya Historical Park

Wat Yai Chai Mongkhon, Ayutthaya Historical Park

Ayutthaya Historical Park

King Ramathibodi I founded the city of Ayutthaya in the middle of the 14th century and over the centuries the city has come to be of great importance. Wat Chai Wattanaram, built in 1630 and extensively restored, is a relic from the reign of King Prasat Thong. In 1357, the foundation stone was laid for Wat Yai Chai Mongkhon with its gallery of sitting Buddhas and Chedis towering into the sky.

Parc historique d'Ayutthaya

Fondé au milieu du XIVᵉ siècle par le roi Ramathibodi I, la ville d'Ayutthaya est devenue au fil des siècles un lieu d'importance majeure. Wat Chai Watthanaram, construit en 1630 et depuis largement restauré, est un vestige du règne du roi Prasat Thong. En 1357 a été érigé le temple de Wat Yai Chai Mongkhon, avec sa galerie de Bouddhas assis et son imposant Cheri.

Ayutthaya-Geschichtspark

König Ramathibodi I. gründete die Stadt Ayutthaya Mitte des 14. Jahrhunderts. Im Laufe der Jahrhunderte gelangte der Ort zu großer Bedeutung. Davon berichtet der 1630 erbaute und aufwendig restaurierte Wat Chai Wattanaram. Er ist ein Relikt aus der Herrschaftszeit von König Prasat Thong. Wenige Im Jahr 1357 legte man den Grundstein für den Wat Yai Chai Mongkhon mit seiner Galerie von sitzenden Buddhas und in den Himmel ragenden Chedis.

Parque Histórico Ayutthaya

El rey Ramathibodi I fundó la ciudad de Ayutthaya a mediados del siglo XIV. Con el paso de los siglos, el lugar ha ido cobrando importancia. El templo Wat Chaiwattanaram, se construyó en 1630 y ha sido objeto de numerosas restauraciones. Constituye una reliquia del reinado del rey Prasat Thong. En 1357, se colocó la primera piedra del monasterio de Wat Yai Chai Mongkhon, con su galería de Budas sentados y chedis elevándose hacia el cielo.

Parco storico di Ayutthaya

Re Ramathibodi I fondò la città di Ayutthaya verso la metà del XIV secolo. Nel corso dei secoli, l'importanza del luogo crebbe. Ne è testimonianza il Wat Chai Wattanaram, edificato nel 1630 sotto il regno di Prasat Thong e oggi restaurato. Nel 1357 erano cominciati i lavori per il Wat Yai Chai Mongkhon, con la sua galleria di Buddha seduti e chedi che si stagliano contro il cielo.

Historisch park Ayutthaya

Koning Ramathibodi I stichtte de stad Ayutthaya in het midden van de 14e eeuw. In de loop der eeuwen is de plaats van steeds groter belang geworden. Dat is bijvoorbeeld te zien aan de in 1630 gebouwde en kostbaar verbouwde Chai Wattanaram. Het is een overblijfsel uit de tijd van koning Prasat Thong. In 1357 werd de eerste steen gelegd voor de tempel Wat Yai Chai Mongkhon, met zijn galerij met zittende boeddha's en hoog oprijzende chedi's.

Ayutthaya Historical Park

Kanchanaburi

Sangkhlaburi

River Kwai

Kanchanaburi

The picturesque province of Kanchanaburi in western Thailand nestles against the border with Myanmar. River landscapes, rainforests, waterfalls and mysterious caves are characteristic of the region, which is only about 100 km from Bangkok. The beauties of nature are in the foreground; no wonder, with a total of seven national parks designated in the province. On the other hand the city of Kanchanaburi, which became famous throughout the world with the Hollywood film *The Bridge on the River Kwai,* is full of hustle and bustle.

Kanchanaburi

La pittoresque province de Kanchanaburi, dans l'ouest de la Thaïlande, est nichée contre la frontière du Myanmar. Paysages fluviaux, forêts tropicales, chutes d'eau et grottes mystérieuses peuplent la région, qui n'est qu'à 100 km de Bangkok. Avec une nature d'une telle beauté, ce n'est pas étonnant que la région comporte sept parcs nationaux. La ville de Kanchanaburi, devenue célèbre dans le monde entier avec le film hollywoodien *Le Pont de la rivière Kwai,* est en revanche pleine d'agitation.

Kanchanaburi

Die malerische Provinz Kanchanaburi im Westen Thailands schmiegt sich an die Grenze von Myanmar. Flusslandschaften, Regenwälder, Wasserfälle und geheimnisvoll erscheinende Höhlen sind charakteristisch für die nur etwa 100 km von Bangkok entfernte Region. Die Schönheiten der Natur stehen im Vordergrund, kein Wunder bei insgesamt sieben in der Provinz ausgewiesenen Nationalparks. Reichlich Trubel herrscht dagegen in der Stadt Kanchanaburi, die durch den Hollywood-Film *Die Brücke am River Kwai* in aller Welt berühmt wurde.

Stalactite cave

Kanchanaburi

La pintoresca provincia de Kanchanaburi, en el oeste de Tailandia, se encuentra junto a la frontera con Myanmar. Los paisajes y bosques fluviales, las cascadas y las misteriosas cuevas caracterizan la región, que se encuentra a sólo unos 100 km de Bangkok. Las bellezas de la naturaleza ocupan un primer plano, algo que no es de extrañar con un total de siete parques nacionales designados en la provincia. La ciudad de Kanchanaburi, por otra parte, que se hizo célebre en todo el mundo con la película de Hollywood *El puente sobre el río Kwai,* es un estallido de vida y bullicio.

Kanchanaburi

La pittoresca provincia di Kanchanaburi, nella Thailandia occidentale, si estende lungo il confine con il Myanmar. Paesaggi fluviali, foreste pluviali, cascate e grotte ricche di mistero sono caratteristici della regione, che dista solo 100 km da Bangkok. Le bellezze della natura stanno qui in primo piano, cosa che non meraviglia vista la presenza di sette parchi nazionali. Fa da contrasto il trambusto che vige nella città di Kanchanaburi, famosa in tutto il mondo per il film americano *Il ponte sul fiume Kwai.*

Kanchanaburi

De schilderachtige provincie Kanchanaburi in het westen van Thailand ligt tegen de grens van Myanmar. Rivierlandschappen, regenwouden, watervallen en mysterieus aandoende grotten zijn karakteristiek voor de regio, die slechts zo'n 100 km van Bangkok ligt. De schoonheden van de natuur staan op de voorgrond – geen wonder in een provincie met in totaal zeven nationale parken. In de stad Kanchanaburi, die wereldberoemd is geworden door de Hollywoodfilm *The Bridge on the River Kwai,* heerst daarentegen een grote bedrijvigheid.

Mon Bridge, Sangkhlaburi

Floating temple, Sangkhlaburi

Wat Wang Wiwekaram, Sangkhlaburi

Spiritual Pilgrimage Sites

The multi-storey pagoda as well as the remaining buildings of the Buddhist Wat Thaworn Wararam have clearly recognizable elements of Chinese style. On the ground floor, fearsome dragon creatures guard the entrance. One of the region's most spiritually important temples is in Sangkhla Buri. Pilgrims worship this place because Wat Wang Wiwekaram is home to the shrine of the revered monk Luang Phaw Uttama, who died in 2006.

Lugares espirituales de peregrinación

La pagoda de varios pisos, así como las construcciones restantes del templo budista Wat Thaworn Wararam, tienen elementos de estilo chino claramente reconocibles. En la planta baja, temibles dragones vigilan la entrada. Uno de los templos espirituales más importantes de la región se encuentra en Sangkhla Buri. Los peregrinos adoran este lugar porque aquí, en el templo Wat Wang Wiwekaram, se puede visitar el santuario del venerado monje Luang Phaw Uttama, que murió en 2006.

Lieux spirituels de pèlerinage

La pagode à plusieurs étages ainsi que les bâtiments restants du Wat Thaworn Wararam bouddhiste comportent des éléments de style chinois clairement reconnaissables. Au rez-de-chaussée, des dragons redoutables gardent l'entrée. L'un des temples les plus importants de la région se trouve à Sangkhla Buri. Les pèlerins adorent cet endroit car il est possible ici de visiter le sanctuaire du vénéré moine Luang Phaw Uttama, décédé en 2006.

Luoghi di pellegrinaggio

La pagoda a più piani e i restanti edifici del tempio buddista Wat Thaworn Wararam presentano evidenti tracce dello stile cinese. Al piano terra, temibili draghi custodiscono l'ingresso. Uno dei templi più importanti della regione si trova a Sangkhla Buri. I pellegrini venerano questo luogo perché qui, a Wat Wang Wiwekaram, si può visitare il santuario del venerato monaco Luang Phaw Uttama, morto nel 2006.

Spirituelle Pilgerstätten

Die mehrstöckige Pagode wie auch die restlichen Gebäudes des buddhistischen Wat Thaworn Wararam weisen klar erkennbar chinesische Stilelemente auf. Im Erdgeschoss bewachen angsteinflößende Drachenwesen den Eingang. Einer der spirituell wichtigsten Tempel der Region steht in Sangkhla Buri. Pilger verehren dieses Ort, weil hier im Wat Wang Wiwekaram der Schrein des hochverehrten Mönchs Luang Phaw Uttama, der 2006 verstarb, besucht werden kann.

Spirituele bedevaartsplaatsen

De pagode met meerdere verdiepingen en de overige gebouwen van het boeddhistische tempelcomplex Wat Thaworn Wararam vertonen duidelijk Chinese stijlelementen. Op de begane grond bewaken angstaanjagende drakenwezens de ingang. Een van de spiritueel belangrijkste tempels van de regio bevindt zich in Sangkhla Buri. Pelgrims aanbidden deze plek omdat hier in Wat Wang Wiwekaram de schrijn van de in 2006 overleden, hooggeachte monnik Luang Phaw Uttama te bezichtigen is.

Khao Chang Phueak Mountain, Thong Pha Phum National Park

Huay Mae Khamin Waterfall, Khuean Si Nakharin National Park

Sai Yok National Park

Around Sai Yok National Park

The popular Sai Yok National Park covers an area of nearly 1000 km². Visitors enjoy numerous caves and impressive waterfalls. Gibbons, elephants, and wild boars live in the teak and deciduous forests and along the rivers, but this protected and mountainous refuge is also home to a tiny, very rare species of bat. The short ride on the notorious Thailand-Burma railway is also adventurous. The line, also known as the Death Railway, was built by forced labourers and prisoners of war recruited during the Second World War by the Japanese army. Many did not survive the inhumane working conditions.

Autour du parc national de Sai Yok

Le populaire parc national de Sai Yok couvre une superficie de près de 1 000 km². Les visiteurs apprécient ses nombreuses grottes et ses chutes d'eau impressionnantes. Gibbons, éléphants et sangliers vivent dans les forêts de teck et de feuillus et le long des rivières, mais ce refuge protégé et montagneux abrite également une minuscule et très rare espèce de chauve-souris. Le court voyage sur le fameux chemin de fer qui relie la Thaïlande et la Birmanie est également une aventure. La ligne, également connue sous le nom de « Death Railway », a été construite pendant la seconde guerre mondiale par l'armée japonaise pour acheminer des travailleurs forcés et des prisonniers de guerre. Beaucoup n'ont pas survécu aux conditions de travail inhumaines.

Rund um den Sai Yok National Park

Der beliebte Sai Yok Nationalpark bedeckt eine Fläche von nahezu 1000 km². Besucher erfreuen sich an zahlreichen Höhlen und imposanten Wasserfällen. In den Teak- und Laubwäldern und an den Flussläufen leben Gibbons, Elefanten, Wildschweine, aber auch eine winzige sehr seltene Fledermausart nennt das geschützte und bergige Refugium ihr zuhause. Ebenfalls abenteuerlich gestaltet sich die kurze Fahrt mit der berüchtigten Thailand-Burma-Eisenbahn. Die auch als „Death Railway" bekannte Linie erbauten während des Zweiten Weltkriegs von der japanischen Armee rekrutierte Zwangsarbeiter und Kriegsgefangene. Viele überlebten die unmenschlichen Arbeitsbedingungen nicht.

Thai - Burma Railway

Alrededor del Parque nacional Sai Yok

El popular parque nacional de Sai Yok abarca un área de casi 1000 km². Los visitantes disfrutan de numerosas cuevas e impresionantes cascadas. Los gibones, elefantes y jabalíes viven en los bosques de teca caducifolios y en las riberas de los ríos, pero este refugio protegido y montañoso es también el hogar de una especie de murciélago diminuta y muy rara. El viaje breve en el célebre ferrocarril entre Tailandia y Birmania también es una aventura. La línea, también conocida como el Ferrocarril de la Muerte, fue construida durante la Segunda Guerra Mundial por el ejército japonés para reclutar trabajadores forzados y prisioneros de guerra. Muchos no sobrevivieron a las inhumanas condiciones de trabajo.

Intorno al Parco nazionale di Sai Yok

Il famoso Parco nazionale di Sai Yok si estende su una superficie di quasi 1.000 km². I visitatori possono ammirare numerose grotte e cascate impressionanti. Gibboni, elefanti e cinghiali vivono nelle foreste di teak e latifoglie e lungo i fiumi, ma questo rifugio protetto e montano ospita anche una piccola specie di pipistrelli molto rara. Avventuroso anche il breve viaggio sulla famosa Thai-Burma Railway. La linea, detta anche "ferrovia della morte", fu costruita durante la seconda guerra mondiale dall'esercito giapponese. Molti degli operai che la realizzarono, costretti ai lavori forzati o prigionieri di guerra, non sono sopravvissuti alle condizioni di lavoro disumane.

Rondom het nationale park Sai Yok

Het populaire nationale park Sai Yok beslaat een oppervlakte van bijna 1000 km². Bezoekers genieten van de talrijke grotten en indrukwekkende watervallen. Gibbons, olifanten en wilde zwijnen leven in de teak- en loofbossen en langs de rivieren, maar dit beschermde en bergachtige toevluchtsoord herbergt ook een kleine, zeer zeldzame vleermuissoort. De korte tocht over de beruchte Birma-Siamspoorweg is ook avontuurlijk. De lijn, die ook bekendstaat als de Dodenspoorlijn, werd tijdens de Tweede Wereldoorlog aangelegd door dwangarbeiders en krijgsgevangenen van het Japanse leger. Velen overleefden de onmenselijke werkomstandigheden niet.

Songkalia River, Sangkhlaburi

Sangkhlaburi

On the border to Myanmar

The mountainous region of Sangkhla Buri with its meandering rivers and dense forests seems almost impenetrable, yet on the Myanmar border live numerous proud hill tribes such as the Hmong, who grow highland rice and soybeans on small areas and have remained largely faithful to their religious and cultural traditions. Between lush vegetation and steep hills, glittering temples appear again and again, like the Buddhist Wat Tha Khanun near the Kwai river. The pagoda, accessible by hundreds of steps, offers a fantastic view over the rugged landscape.

À la frontière du Myanmar

Avec ses rivières sinueuses et ses forêts denses, la région montagneuse de Sangkhla Buri semble presque impénétrable. Pourtant, à la frontière avec le Myanmar, vivent de nombreuses tribus montagnardes comme les Hmong, qui cultivent du riz et du soja sur de petites surfaces et qui sont restés largement fidèles à leurs traditions religieuses et culturelles. Entre végétation luxuriante et collines abruptes, des temples scintillants apparaissent, comme le Wat Tha Khanun bouddhiste, près de la rivière Kwai. La pagode, que l'on peut atteindre en grimpant quelques centaines de marches, offre une vue fantastique sur le paysage accidenté.

An der Grenze zu Myanmar

Das Bergland von Sangkhla Buri mit seinen mäandrierenden Flüssen und dichten Wäldern erscheint nahezu undurchdringlich. Und doch leben hier an der Grenze zu Myanmar zahlreiche stolze Bergvölker wie die Hmong, die auf kleinen Flächen Hochlandreis und Sojabohnen anbauen und ihren religiösen wie kulturellen Traditionen weitgehend treu geblieben sind. Zwischen üppiger Vegetation und steilen Hügel tauchen immer wieder glanzvolle Tempelanlagen auf, wie der buddhistische Wat Tha Khanun in der Nähe des Kwai-Flusses. Von der über hunderte Stufen erreichbaren Pagode zeigt sich ein fantastischer Ausblick über die schroffe Landschaft.

Wat Tha Khanun

En la frontera con Myanmar

La región montañosa de Sangkhla Buri, con sus serpenteantes ríos y densos bosques, parece casi impenetrable. Sin embargo, en la frontera con Myanmar viven numerosas tribus orgullosas de habitar en las montañas, como los Hmong, que cultivan arroz y soja de las tierras altas en pequeñas zonas y se han mantenido fieles en gran medida a sus tradiciones religiosas y culturales. Entre la vegetación exuberante y las colinas empinadas sobre las que brillan complejos religiososaparecen una y otra vez templos como elWat Tha Khanun, de culto budista, cerca del río Kwai. La pagoda, a la que se puede acceder subiendo más de cien escalones, ofrece una vista fantástica sobre el paisaje escarpado.

Al confine con il Myanmar

La regione montuosa di Sangkhla Buri, con i suoi fiumi serpeggianti e le fitte foreste, appare quasi impenetrabile. Tuttavia, qui, al confine con il Myanmar, vivono numerose popolazione montane come quella degli Hmong, che coltivano riso di alta montagna e soia su piccole aree e che sono rimasti in gran parte fedeli alle loro tradizioni religiose e culturali. Tra la vegetazione lussureggiante e le ripide colline compaiono di tanto in tanto lucenti templi come quello buddista di Wat Tha Khanun, vicino al fiume Kwai. La pagoda, a cui si accede salendo centinaia di scalini, offre una vista spettacolare sull'aspro paesaggio.

Aan de grens met Myanmar

Het berglandschap van Sangkhla Buri met zijn meanderende rivieren en dichte bossen lijkt bijna ondoordringbaar. Toch leven hier op de grens met Myanmar tal van trotse bergvolkeren, zoals de Hmong, die op kleine oppervlakken rijst en sojabonen verbouwen en hun religieuze en culturele tradities grotendeels trouw zijn gebleven. Tussen weelderige vegetatie en steile heuvels duiken steeds weer fonkelende tempels op, zoals de boeddhistische Wat Tha Khanun in de buurt van de rivier de Kwai. De pagode, die via honderden traptreden te bereiken is, biedt een fantastisch uitzicht over het ruige landschap.

Wat Tham Sua

Erawan Waterfall, Erawan National Park

Erawan National Park

Erawan National Park is a very popular destination located about 70 km northwest of the city of Kanchanaburi. Between mountain peaks rising up to 996 m and dense deciduous forests, picturesque waterfalls plunge over steep rock edges and create idyllic oases that invite you to a refreshing swim. The most beautiful and impressive is Erawan Falls, which flows over seven cascades; it gave the reserve its name. Besides the rich and colourful birdlife, the park is home to endangered Amman macaques and numerous other mammals such as the Asian elephant.

Parc national d'Erawan

Le parc national d'Erawan est une destination très populaire, située à environ 70 km au nord-ouest de la ville de Kanchanaburi. Entre les sommets montagneux atteignant 996 m et les denses forêts de feuillus, des chutes d'eau pittoresques plongent sur des berges escarpées et créent des oasis idylliques invitant à une baignade rafraîchissante. La plus belle et la plus impressionnante est celle d'Erawan, qui coule en sept cascades et donne son nom à la réserve. Outre une faune aviaire riche et colorée, le parc abrite les macaques amman, en voie de disparition, ainsi que de nombreux autres mammifères comme l'éléphant d'Asie.

Erawan Nationalpark

Als äußerst begehrtes Ausflugsziel gilt der Erawan Nationalpark. Er liegt rund 70 km nordwestlich der Stadt Kanchanaburi. Zwischen bis auf 996 m aufragenden Berggipfeln und dichten Laubwäldern stürzen sich malerische Wasserfälle über steile Felskanten und erschaffen idyllische Oasen, die zu einem erfrischen Bad einladen. Der schönste und imposanteste ist der über sieben Kaskaden dahinfließende Erawan-Wasserfall. Er gab dem Schutzgebiet seinen Namen. Neben einer reichen und bunten Vogelwelt sind im Park die gefährdeten Amman-Makaken und zahlreiche andere Säugetiere wie der Asiatische Elefant anzutreffen.

Assamese Macaque, Erawan National Park

Parque nacional Erawan

El Parque nacional Erawan es un destino muy popular. Se encuentra a unos 70 km al noroeste de la ciudad de Kanchanaburi. Entre montañas de hasta 996 m de altura y densos bosques caducifolios, se precipitan pintorescas cascadas sobre rocas escarpadas creando oasis idílicos que invitan a un refrescante baño. La más bella e impresionante es la cascada de Erawan, que fluye sobre siete cascadas. Le dio su nombre a la reserva. Además de una rica y colorida avifauna, el parque alberga a los amenazados macacos amoníacos y a numerosos mamíferos como el elefante asiático.

Parco nazionale Erawan

Meta molto popolare, il Parco nazionale di Erawan si trova a circa 70 km a nord-ovest della città di Kanchanaburi. Tra vette che arrivano fino a 996 m e fitti boschi di latifoglie, pittoresche cascate scrosciano su rocce scoscese, creando idilliache oasi che invitano a un bagno rinfrescante. La cascata più bella e impressionante è quella di Erawan, che compie ben sette salti e dà il nome al parco. Oltre a un'avifauna ricca e variopinta, il parco ospita i macachi di Amman, una specie in via di estinzione, e numerosi altri mammiferi come l'elefante asiatico.

Nationaal park Erawan

Het nationale park Erawan is een zeer populaire toeristische trekpleister. Het ligt ongeveer 70 km ten noordwesten van de stad Kanchanaburi. Tussen bergtoppen van 996 meter hoog en dichte loofbossen storten pittoreske watervallen over steile rotswanden omlaag en creëren zo idyllische oases die uitnodigen tot een verfrissende duik. De mooiste en indrukwekkendste is de Erawan-waterval, die over zeven cascades omlaag stroomt. Hij gaf het natuurreservaat zijn naam. Naast een gevarieerde en kakelbonte vogelwereld herbergt het park bedreigde Thaise makaken en talloze andere zoogdieren zoals de Aziatische olifant.

Erawan Waterfall, Erawan National Park

Thai Elephant

Bridge over the River Kwai

Bangkok

Skytrain

Chao Phraya River Express boat and Rama VIII bridge

Bangkok

Thailand's capital Bangkok is a vibrant metropolis of millions, with glittering skyscrapers, golden temples and green urban oases. A magical place on the banks of the Chao Phraya. If you want to avoid congested streets on your tour of the city's highlights, jump on the modern metro or take one of the boats that run regularly on the river.

Bangkok

Bangkok, la capitale de la Thaïlande, est une métropole vibrante de plusieurs millions d'habitants, avec des gratte-ciels scintillants, des temples dorés et des oasis urbaines vertes. Les rives du Chao Phraya sont particulièrement magiques. Pour éviter les rues encombrées de la ville, préférez le métro moderne ou les bateaux qui circulent régulièrement sur le fleuve.

Bangkok

Thailands Hauptstadt Bangkok ist eine pulsierende Millionenmetropole mit glitzernden Hochhäusern, goldenen Tempeln und grünen urbanen Oasen. Ein magischer Ort am Ufer des Chao Phraya. Wer auf seiner Tour zu den Highlights der Stadt verstopfte Straßen meiden möchte, steigt in die moderne Metro ein oder nimmt eines der regelmäßig auf dem Fluss verkehrenden Boote.

Bangkok

Bangkok, la capital de Tailandia, es una vibrante metrópolis de millones de habitantes con brillantes rascacielos, templos dorados y verdes oasis urbanos. Un lugar mágico a orillas del río Chao Phraya. Si desea evitar las calles congestionadas en su visita a los lugares más destacados de la ciudad, suba al moderno metro o tome uno de los barcos que circulan regularmente por el río.

Bangkok

Bangkok, la capitale della Thailandia, è una vivace metropoli di milioni di abitanti, con grattacieli scintillanti, templi dorati e oasi verdi nel pieno del paesaggio urbano – un luogo pieno di magia sulle rive del Chao Phraya. Per evitare strade congestionate durante la visita in città, è consigliabile usare la moderna metropolitana o uno dei traghetti che attraversano regolarmente il fiume.

Bangkok

De Thaise hoofdstad Bangkok is een levendige miljoenenstad met glanzende wolkenkrabbers, gouden tempels en groene stadsoases. Een magische plek aan de oevers van de Chao Phraya. Wie op zijn tour naar de hoogtepunten van de stad overvolle straten wil vermijden, stapt in de moderne metro of neemt een van de boten die regelmatig over de rivier varen.

253

Skyline with Skytrain

Grand Palace

Grand Palace – Wat Phra Kaeo

On the grounds of the magnificent palace temple Wat Phra Kaeo, golden beings and sunken Buddha figures appear, apparently immersed in meditation. At the centre of the complex is the temple of the Emerald Buddha, one of the most important and important places of faith of the Thai people. The 66 cm high sitting Buddha figure was carved from a single piece of green jade.

Gran Palacio – Wat Phra Kaeo

En la zona del magnífico templo palaciego aparecen los seres dorados de Wat Phra Kaeo y las figuras de Buda aparentemente hundidas. En el centro del complejo se encuentra el templo del Buda Esmeralda, uno de los lugares de fe más importantes del pueblo tailandés. La figura de Buda sentada, de 66 cm de altura, fue tallada en una sola pieza de jade verde.

Grand Palais – Wat Phra Kaeo

Sur le terrain du magnifique temple du palais Wat Phra Kaeo, apparaissent des statues en or et des statues de Bouddha apparemment en pleine méditation. Au centre du complexe se trouve le temple du Bouddha d'émeraude, l'un des lieux de foi les plus importants du peuple thaïlandais. Le Bouddha assis, de 66 cm de haut, a été sculpté dans un seul morceau de jade vert.

Palazzo Grande – Wat Phra Kaeo

Sul terreno del magnifico tempio-palazzo di Wat Phra Kaeo fanno mostra di sé esseri d'oro e statue del Buddha apparentemente in meditazione. Al centro del complesso si trova il tempio del Buddha di Smeraldo, uno dei più importanti luoghi di fede per il popolo thailandese. La figura del Buddha seduto, alta 66 cm, è stata realizzata da un unico pezzo di giada verde.

Grand Palace – Wat Phra Kaeo

Auf dem Areal des prachtvollen Palasttempels Wat Phra Kaeo zeigen sich goldene Wesen und scheinbar in Mediation versunkene Buddhafiguren. Im Zentrum der Anlage erhebt sich der Tempel des Smaragdbuddhas, einer der wichtigsten und bedeutendsten Glaubensstätten des Thai-Volkes. Die dort untergebrachte rund 66 cm hohe sitzende Buddhafigur wurde aus einem einzigen Stück grüner Jade gehauen.

Grand Palace – Wat Phra Kaeo

Op het terrein van de prachtige paleistempel Wat Phra Kaeo verschijnen gouden wezens en schijnbaar in gedachten verzonken Boeddhabeelden. In het hart van het complex verrijst de tempel van de smaragden Boeddha, een van de belangrijkste heilige plaatsen van het Thaise volk. Het daar ondergebrachte, circa 66 cm hoge zittende Boeddhabeeld werd uit één stuk groene jade gesneden.

Skyline and Chao Phraya River

Buddhist monks, Ratchadamri Road

Wat Phra Dhammakaya

Whoever is on the streets of Bangkok encounters them at every turn: the monks with their orange robes belong to the cityscape in the same way as the rattling three-wheeled tuk-tuks. As a young person, nearly every Thai male goes a to Buddhist temple to spend a few months learning about the simple and religious life of monks. Almost as many Buddha figures as there are monks in Thailand crowd close together in Wat Phra Dhammakaya. Built at the end of the 20th century and continuously extended to this day, the temple is home to around one million golden Buddhas. It is one of the largest Buddhist temples in the world.

Wat Phra Dhammakaya

Qui se balade dans Bangkok les rencontre à chaque coin de ru. Les moines et leurs robes oranges font partie du paysage urbain au même titre que les tuk-tuks à trois roues. Presque tous les jeunes Thaïlandais se rendent quelques mois dans un temple bouddhiste pour connaître la vie simple et religieuse des moines. À Wat Phra Dhammakaya, on trouve presque autant de moines et de statues de Bouddha qu'à la capitale. Construit à la fin du xxᵉ siècle et depuis sans cesse agrandi, le temple abrite environ un million de bouddhas dorés. C'est l'un des plus grands temples bouddhistes du monde.

Wat Phra Dhammakaya

Wer auf den Straßen Bangkoks unterwegs ist, begegnet ihnen auf Schritt und Tritt. Die Mönche mit ihren orangefarbenen Gewändern gehören zum Stadtbild wie die knatternden dreirädrigen Tuk-Tuks. Nahezu jeder Thailänder begibt sich als junger Mensch in einen buddhistischen Tempel, um dort für einige Monate das einfache und religiöse Leben unter Mönchen kennenzulernen. Fast so viele Buddhafiguren wie Mönche in Thailand drängeln sich dicht an dicht im Wat Phra Dhammakaya. Der Ende des 20. Jahrhunderts erbaute und bis heute ständig erweiterte Tempel ist Heimstätte von rund einer Million goldenen Buddhas. Er zählt zu den größten buddhistischen Tempelanlagen der Welt.

Buddha statues, Wat Phra Dhammakaya

Wat Phra Dhammakaya

Quién está en las calles de Bangkok, se encuentra con sus monjes en cada esquina. Estos visten sus túnicas anaranjadas y pertenecen al paisaje de la ciudad como los tuk-tuks de tres ruedas. Casi todos los tailandeses van de jóvenes a un templo budista para conocer la vida sencilla y religiosa de los monjes durante unos meses. En el Wat Phra Dhammakaya se apiñan casi tantas figuras de Buda como monjes en Tailandia. Construido a finales del siglo XX y en continua expansión hasta nuestros días, el templo alberga alrededor de un millón de Budas de oro. Es uno de los templos budistas más grandes del mundo.

Wat Phra Dhammakaya

Per le strade di Bangkok è facile incontrarli ad ogni angolo: i monaci buddisti, con le loro tonache arancioni, fanno parte del paesaggio urbano tanto quanto il piccolo tuk-tuk, il taxi a tre ruote. Quasi ogni thailandese trascorre in gioventù alcuni mesi in un tempio buddista per conoscere la vita semplice e spirituale dei monaci. Nel tempio di Wat Phra Dhammakaya le statue del Buddha sono numerose come i monaci in Thailandia: costruito alla fine del XX secolo e continuamente ampliato fino ad oggi, il tempio buddista, uno dei più grandi al mondo, ospita circa un milione di Buddha d'oro.

Wat Phra Dhammakaya

Wie door de straten van Bangkok loopt, ontmoet ze te pas en te onpas. De monniken met hun oranje gewaden horen bij het stadsbeeld, net als de ratelende driewielige tuktuks. Bijna iedere Thai gaat als jongeling naar een boeddhistische tempel om daar gedurende een paar maanden kennis te maken met het eenvoudige en religieuze leven van monniken. Bijna evenveel Boeddhabeelden als monniken verdringen zich in de Wat Phra Dhammakaya. Deze tempel, die aan het eind van de 20e eeuw gebouwd werd en tot op heden voortdurend wordt uitgebreid, is het onderkomen van ongeveer een miljoen gouden boeddha's. Het is een van de grootste boeddhistische tempelcomplexen ter wereld.

Train Night Market Ratchada

Wat Pho

The lying Buddha

The gilded reclining Buddha statue in the temple Wat Pho, more than 40 m long and about 15 m high, radiates an awe-inspiring atmosphere. Despite the hustle and bustle, one can enjoy the uniqueness of this sacred place and take a look at the finely designed and colourful mural displaying religious scenes. In the hall of the resting Buddha, 108 metal alms bowls are placed, referring to the 108 mother-of-pearl symbols worked into the soles of the feet of the mighty Buddha figure. Throwing coins into every single bowl promises happiness and blessings, and fills the room with a very special sound.

Le Bouddha couché

La statue dorée du Bouddha couché dans le temple Wat Pho, longue de plus de 40 m et haute d'environ 15 m, dégage une atmosphère impressionnante. Malgré l'agitation de ce lieu sacré, il faut en apprécier le caractère unique et jeter un coup d'œil aux peintures murales finement conçues et colorées qui représentent des scènes religieuses. Dans le hall du Bouddha allongé, 108 bols d'aumônes en métal font référence aux 108 symboles de nacre travaillés dans la plante des pieds de la puissante figure du Bouddha. Lancer des pièces de monnaie dans chaque bol promet le bonheur et la bénédiction, et remplit la salle d'un son très spécial.

Der liegende Buddha

Eine ehrfurchtsvolle Atmosphäre strahlt die über 40 m lange und rund 15 m hohe vergoldete, liegende Buddha-Statue im Tempel Wat Pho aus. Trotz des Trubels an diesem Ort sollte man die Einzigartigkeit dieser heiligen Stätte genießen und einen Blick auf die feingestalteten und bunten Wandmalereien mit religiösen Szenen werfen. In der Halle des ruhenden Buddhas sind 108 metallene Almosenschalen aufgestellt, hinweisend auf die 108 in die Fußsohlen der mächtigen Buddhafigur eingearbeiteten Perlmuttsymbole. Das Hineinwerfen von Geldmünzen in jede einzelne Schale verspricht Glück und Segen und erfüllt den Raum mit einem ganz besonderen Klang.

Wat Pho

El Buda yacente

La estatua de Buda recostada en el templo Wat Pho, de más de 40 m de largo y unos 15 m de alto, está bañada en oro e irradia una atmósfera impresionante. A pesar del ajetreo de este lugar, uno debe disfrutar de la singularidad de este lugar sagrado y echar un vistazo a las pinturas murales finamente diseñadas y coloridas con escenas religiosas. En la sala del Buda yacente se colocan 108 tazones metálicos de limosna, refiriéndose a los 108 símbolos de nácar trabajados en las plantas de los pies de la poderosa figura de Buda. Tirar monedas en cada tazón de fuente promete felicidad y bendición y llena la habitación con un sonido muy especial.

Il Buddha sdraiato

La statua dorata del Buddha sdraiato, nel tempio di Wat Pho, lunga più di 40 m e alta circa 15 m, emana grande soggezione. Nonostante il trambusto, ci si dovrebbe abbandonare alla singolarità di questo luogo sacro e osservare con attenzione i variopinti dipinti murali finemente disegnati raffiguranti scene religiose. Nella sala del Buddha sdraiato sono collocate 108 ciotole metalliche per l'elemosina, una per ognuno dei 108 simboli in madreperla inseriti nelle suole dei piedi dell'imponente figura. Gettare monete in ogni singola ciotola porta felicità e benedizione – e riempie la stanza di un suono molto speciale.

De liggende Boeddha

Het vergulde, liggende Boeddhabeeld in de tempel Wat Pho, dat meer dan 40 meter lang is en ongeveer 15 meter hoog, straalt een eerbiedwaardige sfeer uit. Ondanks de mensenmassa's hier zou u moeten genieten van de uniciteit van deze heilige plaats en een blik werpen op de delicaat uitgevoerde en kleurrijke muurschilderingen met religieuze taferelen. In de hal van de rustende Boeddha zijn 108 metalen aalmoezenkommen neergezet, die verwijzen naar de 108 parelmoersymbolen die in de voetzolen van de machtige Boeddha zijn aangebracht. In elke kom een muntstuk werpen belooft geluk en zegen en vult de ruimte met een heel bijzondere klank.

Bangkok

Bangkok's Streets

From a bird's eye view, Bangkok's streets and motorways appear at night like modern, artistically staged light installations. In fact, these are urban masterpieces of modern engineering that are nevertheless unable to cope with the volume of traffic in the Thai capital. Anyone standing in a traffic jam at the Wongwai roundabout can pass the time and marvel at the statue of King Taksin the Great that has been erected in the middle. Its fame is based on the liberation of the ancient capital of Ayutthaya from the Burmese in 1767.

Routes de Bangkok

Vues d'en haut, les rues et les autoroutes de Bangkok peuvent ressembler la nuit à des installations lumineuses modernes et artistiques ; ces chefs-d'œuvre de l'ingénierie urbaine ne sont pourtant pas à la hauteur du volume de trafic qu'abrite la capitale thaïlandaise. Quiconque se trouve dans un embouteillage au rond-point de Wongwai pourra tout de même passer le temps en s'émerveillant devant la statue qui trône en son milieu. Elle représente le roi Taksin le Grand, célèbre pour avoir participé à la libération de l'ancienne capitale d'Ayutthaya des mains des Birmans, en 1767.

Bangkoks Straßen

Aus der Vogelperspektive erscheinen Bangkoks Straßen und Autobahnen bei Nacht wie moderne, kunstvoll inszenierte Lichtinstallationen. Tatsächlich handelt es sich um urbane Meisterstücke moderner Ingenieurskunst, die dem Verkehrsaufkommen in der Thailändischen Hauptstadt dennoch nicht gewachsen sind. Wer am Wongwai-Kreisverkehr im Stau steht, kann sich die Zeit vertreiben und die in der Mitte errichtete Statue von König Taksin dem Großen bestaunen. Sein Ruhm beruht auf der Befreiung der alten Hauptstadt Ayutthayas von den Birmesen im Jahr 1767.

Wongwai 22 Roundabout

Calles de Bangkok

A vista de pájaro, las calles y autopistas de Bangkok aparecen de noche como modernas instalaciones de iluminación artísticamente escenificadas. De hecho, se trata de obras maestras del arte moderno de la ingeniería urbana que, sin embargo, no están a la altura del volumen de tráfico de la capital tailandesa. Cualquiera que se encuentre en un embotellamiento en la rotonda de Wongwai, puede contemplar y maravillarse con la estatua del rey Taksin el Grande, erigida en el centro, conocido por la liberación de la antigua capital de Ayutthaya de los birmanos, en 1767.

Le strade di Bangkok

Se viste dall'alto di notte, le strade e le autostrade di Bangkok appaiono come moderne installazioni luminose, assemblate in maniera decisamente artistica. In realtà, si tratta di capolavori urbani di moderna ingegneristica che tuttavia non sono all'altezza del traffico della capitale thailandese. Chi si trovi bloccato in un ingorgo sulla rotatoria di Wongwai può passare il tempo ammirando la statua del re Taksin il Grande, eretta al centro dell'incrocio. La sua fama è dovuta alla liberazione dell'antica capitale di Ayutthaya nel 1767, che si trovava allora in mano birmana.

Straten van Bangkok

Vanuit vogelperspectief zien de straten en snelwegen van Bangkok er 's nachts uit als moderne, artistiek geënsceneerde lichtinstallaties. Het zijn in feite stedelijke meesterwerken van moderne ingenieurskunst die niet opgewassen zijn tegen het drukke verkeer in de Thaise hoofdstad. Wie op de rotonde van Wongwai in de file staat, kan zijn tijd verdrijven door het standbeeld van koning Taksin de Grote te bewonderen dat in het midden staat. Diens roem is gebaseerd op de bevrijding van de oude hoofdstad Ayutthaya op de Birmezen in 1767.

Chao Phraya River

Tuk-tuk

On the road in Bangkok

With the year-round tropical temperatures, Bangkok traders are happy to locate their colourful sales stands outside. The night markets are popular with locals and visitors alike. The offerings are varied, ranging from T-shirts to spicy Thai curries. The best and cheapest way to reach such markets or food stalls is the tuk-tuk.

Por Bangkok

Con temperaturas tropicales durante todo el año, los vendedores de Bangkok se complacen en trasladar sus coloridos puestos de venta al exterior. Los mercados nocturnos son populares tanto entre los locales como entre los visitantes. La oferta es variada. Se vende desde camisetas hasta curry tailandés picante. La mejor manera y la más barata de llegar a estos mercados o cocinas es el tuk-tuk.

Sur la route à Bangkok

Avec des températures tropicales toute l'année, les marchands de Bangkok sont heureux de déplacer leurs stands de vente colorés à l'extérieur. Les marchés nocturnes sont populaires auprès de la population locale comme des visiteurs. L'offre est variée et va du T-shirt au curry thaïlandais épicé. Le meilleur moyen – et le moins cher – d'y accéder est le tuk-tuk.

In giro per Bangkok

Con temperature tropicali tutto l'anno, i mercanti di Bangkok sono più che felici di portare i loro stand variopinti all'aperto. I mercati notturni sono molto popolari sia tra la gente del posto che tra i turisti. L'offerta è varia e spazia dalle T-shirt ai curry thailandesi piccanti. Il modo migliore e più economico per raggiungere tali mercati e i ristorantini sulla strada è il tuk-tuk.

Unterwegs in Bangkok

Bei ganzjährig tropischen Temperaturen verlagern die Bangkoker Händler ihre bunten Verkaufsstände gerne ins Freie. Die Nachtmärkte sind bei Einheimischen und Besuchern gleichermaßen beliebt. Das Angebot ist vielfältig und reicht vom T-Shirt bis zum scharf gewürzten Thai-Curry. Das beste und günstigste Verkehrsmittel, um solche Märkte oder Garküchen zu erreichen, ist das Tuk-Tuk.

Onderweg in Bangkok

Met tropische temperaturen het hele jaar door verplaatsen handelaars in Bangkok hun kakelbonte marktkramen graag naar buiten. De avondmarkten zijn zowel bij de lokale bevolking als bij bezoekers erg geliefd. Het aanbod is zeer divers en varieert van T-shirts tot pittige Thaise curry's. De beste en goedkoopste manier om dergelijke markten of eetkramen te bereiken is per tuk-tuk.

Chinese New Year, Chinatown

Chinatown

The Chinese Quarter around Yaowarat Road embodies
its own, self-contained world. Chinese characters,
wherever you look. Here, the often long-established
Chinese celebrate their traditional festivals such as
the New Year's festival, which is always celebrated
on a new moon day between January and February.
Bustling streets, family-run shops, small hotels and
lively markets also characterise this busy area.

Quartier chinois

Le quartier chinois autour de Yaowarat Road est
presque un monde autonome, et où que vous
regardiez, vous y trouverez des caractères chinois.
Ici, les habitants, souvent établis de longue date,
célèbrent leurs fêtes traditionnelles comme la fête
du Nouvel An, qui a toujours lieu un jour de nouvelle
lune entre janvier et février. Les rues, les boutiques
familiales, les petits hôtels et les marchés font vivre ce
quartier animé.

Chinatown

Eine eigene und in sich abgeschlossene Welt
verkörpert das China-Viertel rund um die Yaowarat
Road. Chinesische Schriftzeichen, wo hin man schaut.
Hier feiern die oft alteingesessen Chinesen ihre
traditionellen Feste wie das Neujahrfest, das immer
an einem Neumondtag zwischen Januar und Februar
zelebriert wird. Außerdem prägen geschäftige
Straßen, familiengeführte Ladenlokale, kleine Hotels
und quirlige Märkte das Bild dieser ruhelosen Gegend.

Yaowarat Road, Chinatown

Barrio Chino

El barrio chino alrededor de Yaowarat Road encarna su propio mundo autónomo. Caracteres chinos, dondequiera que mires. Los chinos, a menudo establecidos desde hace mucho tiempo, celebran aquí sus fiestas tradicionales, como la fiesta de Año Nuevo, que siempre se celebra en un día de luna nueva entre enero y febrero. Calles bulliciosas, tiendas familiares, pequeños hoteles y mercados animados también caracterizan esta zona inquieta.

Chinatown

Il quartiere cinese che si snoda intorno alla Yaowarat Road è un mondo tutto a sé. Qui campeggiano i caratteri cinesi ovunque si guardi. I cinesi che risiedono qui, spesso anche da molto, festeggiano le loro feste tradizionali come il capodanno, che si celebra tra gennaio e febbraio in un giorno di luna nuova. Strade animate, negozi a conduzione familiare, piccoli alberghi e vivaci mercati sono caratteristici di questa zona.

Chinatown

De Chinese wijk rond Yaowarat Road belichaamt een heel eigen, volledig afgezonderde wereld. Chinese karakters, waar u ook kijkt. Hier vieren de vaak al lang in Thailand woonachtige Chinezen hun traditionele feesten zoals het Nieuwjaarsfeest, dat altijd op een dag met nieuwemaan tussen januari en februari gevierd wordt. Ook hectische straatjes, familiebedrijven, kleine hotels en levendige markten bepalen het beeld van deze onrustige wijk.

Chao Phraya River

CULINARY

Thai cuisine is a melting pot full of exotic scents and flavours. Whether in a food stall on a busy street or in an exclusive restaurant, the almost always freshly prepared dishes are an experience for the senses. Unknown fruits or insects included. The motto is: Be brave, try and enjoy.

GASTRONOMIE

La gastronomie thaïlandaise est un creuset de parfums et de saveurs exotiques. Que ce soit dans la cuisine de rue, animée, ou dans les restaurants, les plats presque toujours fraîchement préparés sont une expérience pour les sens. Fruits ou insectes inconnus inclus – la devise est « Soyez courageux, essayez et appréciez. »

KULINARISCHES

Die thailändische Küche ist ein Schmelztiegel voller exotischer Düfte und Geschmacksrichtungen. Ob in der Garküche an der belebten Straße oder im exklusiven Restaurant, die fast immer frisch zu bereiteten Speisen sind ein Erlebnis für die Sinne. Unbekannte Früchte oder Insekten inklusive: Das Motto lautet: Mutig sein, Ausprobieren und Genießen.

GASTRONOMÍA

La cocina tailandesa es un crisol lleno de aromas y sabores exóticos. Ya sea en la puesto de una concurrida calle o en un exclusivo restaurante, los platos, casi siempre recién preparados, son una experiencia para los sentidos. Frutas o insectos desconocidos incluidos; el lema es "sé valiente, prueba y disfruta".

GASTRONOMIA

La cucina thailandese è un crogiolo di profumi e sapori esotici. Sia nelle cucine improvvisate lungo le strade che nei ristoranti più esclusivi, i piatti, preparati quasi sempre al momento, sono un'esperienza per i sensi. Frutti o insetti sconosciuti inclusi – il motto è: trovare il coraggio, provare e divertirsi.

CULINAIR

De Thaise keuken is een smeltkroes vol exotische geuren en smaken. Of het nu in de eettentjes in drukke straten is of in een exclusief restaurant, de bijna altijd versbereide gerechten zijn een genot voor de zintuigen. Inclusief onbekende vruchten en insecten. Het motto is: wees dapper, proef en geniet.

CATFISH

STREETFOOD

MANGOSTEEN

CHILI

PREPARATION OF STICKY RICE

ROASTED BANANA

MARKET

EGGS

FISH

CHILI SAUCE

FRIED GRASSHOPPERS AND LARVA

SWEET POTATOES

STREETFOOD

PITAYA

COCONUT ICE CREAM

SEAFOOD

285

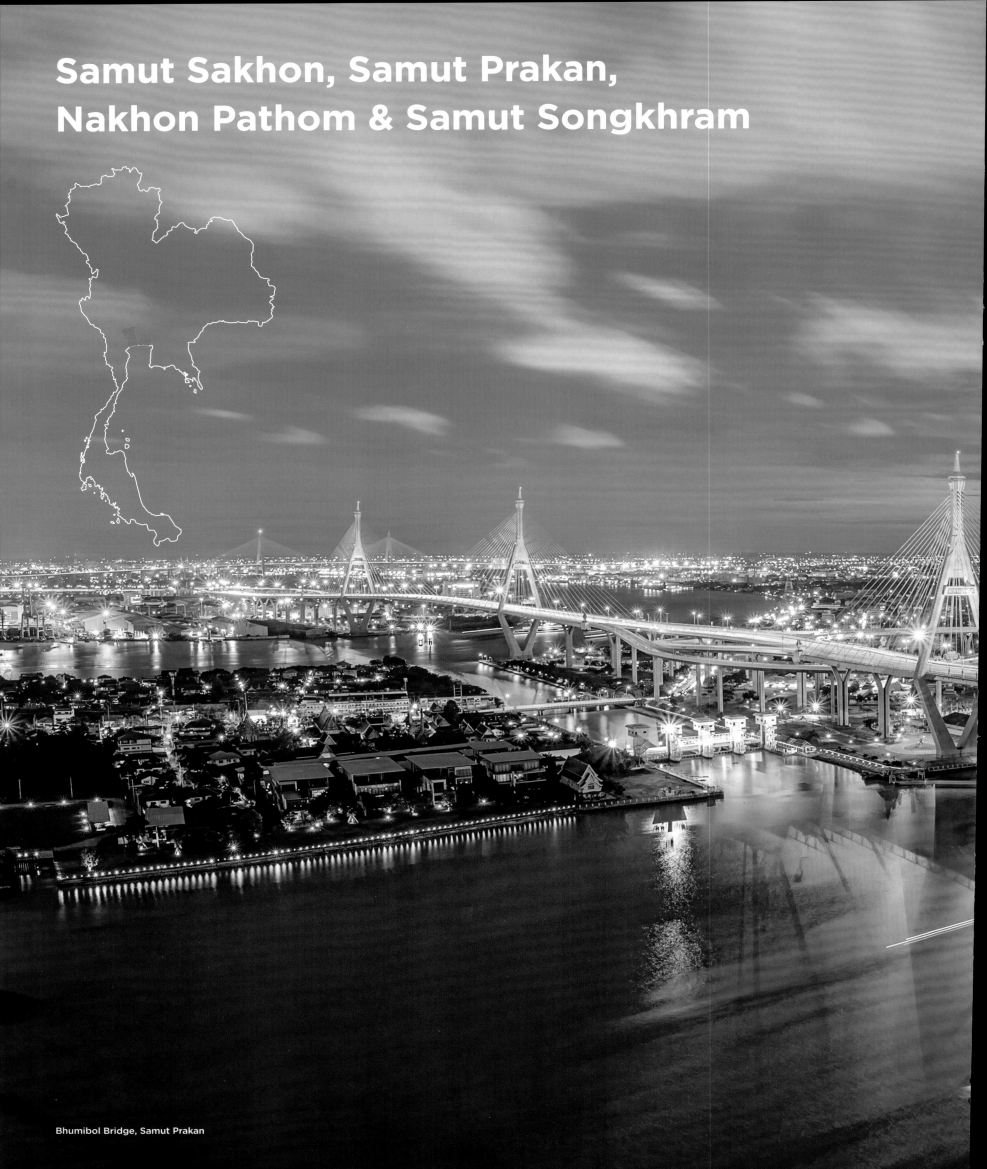

Samut Sakhon, Samut Prakan, Nakhon Pathom & Samut Songkhram

Bhumibol Bridge, Samut Prakan

Ancient City Muang Boran, Samut Prakan

Samut Sakhon, Samut Prakan, Nakhon Pathom & Samut Songkhram

The provinces of Samut Sakhon, Samut Prakan, Nakhon Pathom and Samut Songkhram, bordering Bangkok, unite on the Gulf of Thailand. Here, glittering modernity meets the traditional past. While the fascinating Bhumibol Bridge stretching across the Chao Praya River symbolises progress and new beginnings, intricately decorated temples and Buddha figures represent the living religion of the Thai people.

Samut Sakhon, Samut Prakan, Nakhon Pathom & Samut Songkhram

En el Golfo de Tailandia, en la frontera con Bangkok, se unen las provincias de Samut Sakhon, Samut Prakan, Nakhon Pathom y Samut Songkhram. La brillante modernidad se encuentra en ellas con el pasado tradicional. Mientras que el puente Bhumibol, que se extiende a lo largo del río Chao Praya, simboliza el progreso y la partida, los templos decorados con filigranas y las figuras de Buda representan la religión vivida por los tailandeses.

Samut Sakhon, Samut Prakan, Nakhon Pathom & Samut Songkhram

Sur le golfe de Thaïlande, à la frontière de Bangkok, les provinces de Samut Sakhon, Samut Prakan, Nakhon Pathom et Samut Songkhram se rejoignent, et la modernité scintillante rencontre le passé traditionnel : alors que le pont de Bhumibol, qui s'étend sur la rivière Chao Praya, symbolise le progrès, les temples décorés en filigrane et les statues de Bouddha représentent la religion vivante du peuple thaïlandais.

Samut Sakhon, Samut Prakan, Nakhon Pathom & Samut Songkhram

Nel golfo di Thailandia, nei pressi di Bangkok, si trovano le province di Samut Sakhon, Samut Prakan, Nakhon Pathom e Samut Songkhram. Qui la modernità più sfarzosa incontra il passato ricco di tradizione. Mentre il ponte di Bhumibol, che si allunga sopra il fiume Chao Praya, simboleggia il progresso e lo sviluppo, templi decorati in filigrana e statue del Buddha rappresentano la vita spirituale molto presente nella quotidianità thailandese.

Samut Sakhon, Samut Prakan & Nakhon Pathom

Am Golf von Thailand, angrenzend an Bangkok vereinen sich die Provinzen Samut Sakhon, Samut Prakan, Nakhon Pathom und Samut Songkhram. Dabei trifft die glitzernde Moderne auf traditionsreiche Vergangenheit. Während die sich über den Chao-Phraya-Fluss spannende Bhumibol-Brücke Fortschritt und Aufbruch symbolisiert, stehen filigran verzierte Tempel und Buddhafiguren für die gelebte Religion der Thailänder.

Samut Sakhon, Samut Prakan, Nakhon Pathom & Samut Songkhram

Aan de Golf van Thailand, grenzend aan Bangkok, komen de provincies Samut Sakhon, Samut Prakan, Nakhon Pathom en Samut Songkhram samen. De glinsterende moderniteit stuit daar op het traditionele verleden. Terwijl de Bhumibolbrug, die de rivier de Chao Praya overspant, vooruitgang en verandering symboliseert, staan met filigraan versierde tempels en Boeddhabeelden voor het aangehangen geloof van de Thai.

Ampawa, Samut Songkhram

Exceptional markets

The small town of Ampawa is popular for its floating market and the restaurants and shops on the shore. On weekends it can become especially crowded when the inhabitants of Bangkok pay a brief visit to the village for a quick getaway. Not far from Thailand's capital, the city of Samut Songkhram is also home to a unique railway market. It is an exciting experience, almost a spectacle when the train rolls by a few centimetres from the stalls. As a warning, a loud signal sounds shortly before the train goes by.

Des marchés exceptionnels

La petite ville d'Ampawa est une destination populaire en raison de son marché flottant et de ses restaurants et boutiques sur le rivage. Les week-ends, il peut y avoir beaucoup de monde car les habitants de Bangkok y passent parfois de courtes vacances. Non loin de la capitale thaïlandaise, la ville de Samut Songkhram abrite également un marché ferroviaire unique. Expérience passionnante proche du spectacle, le train frôle les stalles de quelques centimètres, avertissant de son passage que peu de temps avant, à l'aide d'une forte sirène.

Außergewöhnliche Märkte

Die Kleinstadt Ampawa wird wegen ihres schwimmenden Marktes und der am Ufer angrenzenden Restaurants und Geschäften gerne besucht. Am Wochenende kann es besonders voll werden, wenn die Bewohner Bangkoks für einen Kurzurlaub dem Ort eine Stippvisite abstatten. Ebenfalls nicht weit entfernt von Thailands Hauptstadt stößt man in der Stadt Samut Songkhram auf den einzigartigen, sogenannten Eisenbahnmarkt. Es ist eine aufregende Erfahrung, geradezu ein Spektakel, wenn der Zug wenige Zentimeter an den Verkaufsständen vorbeiholpert. Zur Warnung ertönt kurz vor der Durchfahrt eine laute Sirene.

Maeklong Railway Market, Samut Songkhram

Mercados excepcionales

La pequeña ciudad de Ampawa es un destino popular debido a su mercado flotante y restaurantes y tiendas en la costa. Los fines de semana puede haber mucha gente cuando los habitantes de Bangkok visitan el pueblo para pasar unas breves vacaciones. No lejos de la capital de Tailandia, la ciudad de Samut Songkhram también alberga el singular mercado ferroviario. Es una experiencia emocionante, casi un espectáculo cuando el tren pasa por los puestos por unos pocos centímetros. Como advertencia, una sirena fuerte suena poco antes del pasaje.

Mercati insoliti

La piccola città di Ampawa è popolare per il mercato galleggiante, i ristoranti e i negozi sulla riva del fiume. Nei fine settimana il posto può diventare particolarmente affollato quando gli abitanti di Bangkok fanno tappa qui per una breve vacanza. Non lontano dalla capitale, la città di Samut Songkhram è sede di un altro mercato molto particolare, quello detto "della ferrovia": vedere il treno che passa a pochi centimetri dagli stand è un'esperienza emozionante e un vero e proprio spettacolo. Una rumorosa sirena avverte del passaggio imminente.

Buitengewone markten

Het kleine stadje Ampawa is een populaire bestemming vanwege de drijvende markt en de restaurants en winkels langs de oever. Vooral in het weekend kan het erg druk worden, als de inwoners van Bangkok een kort bezoek brengen aan het dorp. Ook niet ver bij de hoofdstad van Thailand vandaan stuit je in de stad Samut Songkhram op de unieke, zogenaamde spoorwegmarkt. Het is een spannende ervaring, bijna een spektakel, als de trein op een paar centimeter afstand langs de kraampjes voorbijhobbelt. Als waarschuwing klinkt kort voor het passeren een luide sirene.

Samut Sakhon

Chonburi

Pattaya City

Chonburi

The provincial capital of Chonburi, less than 100 km from the centre of Bangkok, has hardly any foreign visitors, but almost everyone has heard of Pattaya, the internationally renowned tourist centre on the Gulf of Thailand. Life pulsates here day and night. There is hardly any less activity on Pratumnak Mountain in the middle of the city. The 18 m high Buddha figure there is one of the most popular sights of the seaside resort. A visit to Koh Si Chang Island is more peaceful and relaxing, with some beautiful beaches and great viewpoints over the calm sea and the landscape.

Chonburi

Si la capitale de la province de Chonburi – à laquelle elle donne son nom –, à moins de 100 km du centre de Bangkok, est négligée par les visiteurs étrangers, quasiment tout le monde a déjà entendu parler de Pattaya, la station balnéaire de renommée internationale sur le golfe de Thaïlande. La vie bat ici jour et nuit. Il n'y a guère moins d'activité sur le mont Pratumnak, au milieu de la ville : le Bouddha de 18 m de haut est l'une des curiosités les plus populaires. L'île de Koh Si Chang est plus paisible et charmante, avec de belles plages et de superbes points de vue sur la mer calme et le paysage environnant.

Chonburi

Die knapp 100 km vom Bangkoker Zentrum entfernte Provinzhauptstadt Chonburi kennt kaum ein ausländischer Besucher, dafür hat fast jeder schon einmal von Pattaya gehört, der international bekannten Touristenhochburg am Golf von Thailand. Hier pulsiert das Leben, Tag und Nacht. Kaum weniger Betrieb herrscht auf dem mitten in der Stadt liegenden Pratumnak-Berg. Die dortige 18 m hohe Buddhafigur ist eine der beliebtesten Sehenswürdigkeiten des Badeortes. Geruhsamer und charmanter gestaltet sich der Besuch der Insel Koh Si Chang mit ein paar hübschen Stränden und tollen Aussichtspunkten über das ruhige Meer und die Landschaft.

Big Buddha, Pattaya

Chonburi

La capital de la provincia, Chonburi, a menos de 100 km del centro de Bangkok, apenas conoce a ningún visitante extranjero, pero casi todo el mundo ha oído hablar de Pattaya, la fortaleza turística de renombre internacional en el Golfo de Tailandia. La vida late aquí día y noche. La ciudad aledaña a la montaña Pratumnak también tiene mucha vida. La figura de Buda, de 18 m de altura, es una de las vistas más populares de la estación balnearia. La visita a la isla de Koh Si Chang es más tranquila y cautivadora, con algunas playas bellas y magníficos miradores sobre el mar tranquilo y el paisaje.

Chonburi

La capitale della provincia, Chonburi, a meno di 100 km da Bangkok, conosce appena visitatori stranieri, ma quasi tutti hanno sentito parlare di Pattaya, la roccaforte turistica di fama internazionale sul golfo di Thailandia. Qui la vita pulsa giorno e notte. Ugualmente vivace il monte Pratumnak, nel centro della città. Il Buddha di 18 m di altezza è una delle attrazioni più popolari di questa località balneare. Decisamente più tranquilla e affascinante è una visita all'isola di Koh Si Chang, con le belle spiagge e gli ottimi punti di vista sul mare calmo e sul paesaggio circostante.

Chonburi

De op minder dan 100 km van het centrum van Bangkok gelegen provinciehoofdstad Chonburi trekt bijna geen buitenlandse bezoekers, terwijl de meeste mensen weleens hebben gehoord van Pattaya, het internationaal bekende toeristische bolwerk aan de Golf van Thailand. Het leven bruist hier dag en nacht. Nauwelijks minder drukte heerst er op de Pratumnak-berg midden in de stad. Het 18 meter hoge Boeddhabeeld dat daar staat, is een van de geliefdste bezienswaardigheden van de badplaats. Een bezoek aan het eiland Koh Si Chang, met een aantal fraaie stranden en geweldige uitzichtpunten over de kalme zee en het landschap, is gezapiger en charmanter.

Koh Si Chang

Wat Djittabhawan, Pattaya

Pattaya City

Nong Nooch Tropical Botanical Garden, Pattaya

Prasat Sut Ja-Tum (Sanctuary of Truth), Pattaya

Prasat Sut Ja-Tum

Built of wood, the "Sanctuary of Truth" has a proud height of over 100 m and is unique in the world. An almost endless number of finely carved figures and picture scenes adorn this architectural masterpiece standing directly by the sea. Motifs from Buddhism and Hinduism and works of art from many other Asian cultures are presented.

Prasat Sut Ja-Tum

Construit en bois et haut de plus de 100 m, le « Sanctuaire de la vérité » est unique au monde. Un nombre presque infini de statues finement sculptées et de représentations religieuses ornent ce chef-d'œuvre architectural, situé directement au bord de la mer. On peut y observer des motifs issus du bouddhisme et de l'hindouisme, mais aussi des œuvres d'art venues de nombreuses autres cultures asiatiques.

Prasat Sut Ja-Tum

Das aus Holz erbaute „Heiligtum der Wahrheit" hat eine stolze Höhe von über 100 m und ist einmalig auf der Welt. Eine schier endlose Anzahl von fein geschnitzten Figuren und Bilderszenen schmücken das direkt am Meer stehende architektonische Meisterwerk. Dargestellt werden Motive aus dem Buddhismus und Hinduismus und Kunstwerke aus vielen anderen asiatischen Kulturen.

Prasat Sut Ja-Tum

Construido en madera, el "Santuario de la Verdad" tiene una altura de más de 100 m y es único en el mundo. Un número casi infinito de figuras finamente talladas y escenas pictóricas adornan esta obra maestra arquitectónica situada directamente junto al mar. Se presentan motivos del budismo y el hinduismo y obras de arte de muchas otras culturas asiáticas.

Prasat Sut Ja-Tum

Costruito in legno, il "santuario della verità" ha un'altezza di oltre 100 m ed è unico nel suo genere. Un numero quasi infinito di figure finemente scolpite e immagini con varie scene adornano questo capolavoro architettonico che si erge sul mare. Le raffigurazioni rappresentano motivi del buddismo e dell'induismo e opere d'arte di molte altre culture asiatiche.

Prasat Sut Ja-Tum

Het van hout gemaakte "Heiligdom van de waarheid" heeft een trotse hoogte van meer dan 100 meter en is uniek in de wereld. Een bijna eindeloos aantal figuren en beeldtaferelen van het fijnste houtsnijwerk sieren dit architectonische meesterwerk dat direct aan zee staat. Er zijn motieven uit het boeddhisme en hindoeïsme en kunstwerken uit allerlei andere Aziatische culturen afgebeeld.

Beach near Bang Saray

Ban Bang Phra Reservoir, Si Racha

Ban Bang Phra Reservoir, Si Racha

Koh Kham, Sattahip

Koh Kham, Sattahip

Koh Kham, Sattahip

Koh Kham

This tiny island off the coast near the city of Sattahip, about 40 km south of Pattaya, has beautiful beaches and turquoise, crystal-clear water. A picturesque place to relax or to take a leisurely walk under the azure blue sky of Thailand. In the evening it is quiet again on the enchanting island, because you are not allowed to spend the night here.

Koh Kham

Esta pequeña isla frente a la costa de la ciudad de Sattahip, a unos 40 km al sur de Pattaya, tiene hermosas playas y aguas cristalinas de color turquesa. Un lugar pintoresco para relajarse o dar un tranquilo paseo bajo el cielo azul celeste de Tailandia. Por la noche, la isla fascinante se vuelve tranquila, ya que no se nos permite pasar la noche aquí.

Koh Kham

Cette petite île au large de la côte de la ville de Sattahip, à environ 40 km au sud de Pattaya, possède de belles plages et une eau turquoise et cristalline. Un endroit pittoresque pour se détendre ou se promener tranquillement sous le ciel bleu azur de la Thaïlande. Le soir, le calme revient sur l'île enchantée ; en effet, les visiteurs ne sont pas autorisés à y passer la nuit.

Koh Kham

Questa piccola isola al largo della costa della città di Sattahip, circa 40 km a sud di Pattaya, ha belle spiagge e acque turchesi cristalline. Un posto ameno per rilassarsi o per fare piacevoli passeggiate sotto il cielo azzurro della Thailandia. La sera tutto torna ad essere silenzioso su quest'isola incantevole, poiché non è permesso pernottare qui.

Koh Kham

Traumhafte Strände und türkisblaues, glasklares Wasser nennt diese winzige Insel vor der Küste der Stadt Sattahip, etwa 40 km südlich von Pattaya, ihr Eigen. Ein malerischer Ort, zum Entspannen oder um gemütlich einen Spaziergang unter dem azurblauen Himmel Thailands in Angriff zu nehmen. Abends wird es dann wieder ganz still auf dem bezaubernden Eiland, denn übernachtet werden darf hier nicht.

Koh Kham

Dit kleine eilandje voor de kust van de stad Sattahip, ongeveer 40 km ten zuiden van Pattaya, heeft prachtige stranden en turquoise, kristalhelder water. Een schilderachtige plek om te ontspannen of om een gemoedelijke wandeling te maken onder de azuurblauwe hemel van Thailand. 's Avonds is het dan weer volkomen stil op het betoverende eiland, omdat je hier niet mag overnachten.

Rayong

Rayong

Wooden bridge, Rayong

Rayong

Long beaches and numerous beautiful islands off the coast guarantee the popularity of the region around Rayong. Koh Samet is number one on the popularity scale of the islands: a wonderful place for people who like to have a lively time. Those who prefer to include pristine and almost untouched nature on their wish list are in perfect hands on the small Koh Mun islands, secluded beaches included. Culinary delicacies are also on the menu in Rayong. It is said that the seafood here is particularly fresh and delicious.

Rayong

Les longues plages et les nombreuses îles au large de la côte participent à la popularité de la région autour de Rayong – Koh Samet étant probablement tout en haut l'échelle, surtout pour les visiteurs friands d'animation. Ceux qui adorent la nature originale et presque intacte lui préféreront les petites îles Koh Mun et leurs plages désertes. Les délices culinaires sont également au menu à Rayong : on dit que les fruits de mer y sont particulièrement frais et délicieux.

Rayong

Lange Strände und zahlreiche der Küste vorgelagerte Inseln mit Esprit und Grazie sind Garanten für die Beliebtheit der Region um Rayong. Koh Samet steht auf Platz 1 der Beliebtheitsskala der hiesigen Eilande. Ein herrlicher Ort für Menschen, die es gerne lebhaft mögen. Wer lieber ursprüngliche und nahezu unberührte Natur auf seinem Wunschzettel notiert hat, ist auf den kleinen Koh Mun-Inseln perfekt aufgehoben. Einsame Strände inklusive. Ferner stehen in Rayong kulinarische Köstlichkeiten auf dem Speiseplan. Man erzählt sich, dass die Meeresfrüchte hier besonders frisch und ausnehmend lecker seien.

Koh Mun

Rayong

Largas playas y numerosas islas de carácter y gracia frente a la costa garantizan la popularidad de la región alrededor de Rayong. Koh Samet es la primera de las islas en términos depopularidad. Un lugar maravilloso para la gente a la que le gusta la vida. Aquellos que prefieren la naturaleza original y casi virgen, están en han llegado al sitio adecuado, en las pequeñas islas Koh Mun. Playas solitarias incluidas. Las delicias culinarias también están en el menú de Rayong. Se dice que los mariscos aquí son particularmente frescos y deliciosos.

Rayong

Le lunghe spiagge e le numerose isole, adagiate graziosamente davanti alla costa, garantiscono la popolarità all'omonima regione intorno a Rayong. Koh Samet è l'isola più conosciuta, il luogo ideale per chi ama i luoghi vivaci. L'arcipelago di Koh Mun, invece, è più indicato per chi preferisce la natura incontaminata – spiagge solitarie incluse. Anche varie prelibatezze culinarie sono di casa nel Rayong: si dice che qui i frutti di mare siano particolarmente freschi e deliziosi.

Rayong

Lange stranden en talrijke gracieuze eilanden met esprit voor de kust staan garant voor de populariteit van de regio rond Rayong. Koh Samet staat op nummer 1 op de populariteitsschaal van de eilanden alhier. Een heerlijke plek voor mensen die van het leven genieten. Wie oorspronkelijke en bijna ongerepte natuur op zijn verlanglijstje heeft staan, is op de kleine Koh Mun-eilanden perfect op zijn plaats. Eenzame stranden inbegrepen. Verder staan culinaire lekkernijen op het menu in Rayong. Er wordt gezegd dat de vis hier bijzonder vers en buitengewoon lekker is.

Mangrove Forests

Tung Prong Thong, The Golden Mangrove Field

Koh Samet

Koh Samet

Divers love the varied underwater world with its coral gardens, and sun-seekers make a pilgrimage to Koh Samet to relax on the sandy beaches lined with evergreen cajeput trees. The island enchants with its charming bays, and offer good water sports opportunities. Almost all ferries from the mainland arrive in the capital Na Dan. Here is most of the hustle and bustle, and the density of restaurants and bars is comparatively high. On the other hand, the interior of the island is part of the Khao Laem Ya – Mu Ko Samet National Park, and is largely untouched. Many exotic plants and animals can be found there.

Koh Samet

Les plongeurs y apprécieront le monde sous-marin varié et ses barrières de corail tandis que les amateurs de soleil choisiront Koh Samet pour se détendre sur ses plages de sable fin bordées d'arbres à feuillage persistant. L'île enchante en effet par ses baies charmantes et son offre de sports nautiques. Dans la capitale Na Dan, presque tous les ferries arrivent du continent, concentrant une grande partie de l'animation ; la densité des restaurants et des bars y est d'ailleurs relativement élevée. L'intérieur de l'île, appartenant au parc national de Khao Laem Ya - Mu Ko Samet, est en grande partie intact. On y trouve de nombreuses plantes et animaux exotiques.

Koh Samet

Taucher lieben die abwechslungsreiche Unterwasserwelt mit ihren Korallengärten und Sonnenhungrige pilgern nach Koh Samet, um an den mit immergrünen Kajeputbäumen gesäumten Sandstränden zu relaxen. Die Insel bezaubert durch reizende Buchten und ein gutes Angebot an Wassersportmöglichkeiten. Im Hauptort Na Dan kommen fast alle Fähren vom Festland an. Hier herrscht der meiste Trubel und die Dichte an Restaurants und Bars ist vergleichsweise hoch. Weitgehend unberührt präsentiert sich dagegen das Landesinnere der zum Nationalpark Khao Laem Ya – Mu Ko Samet gehörenden Insel. Zu Fuß lassen sich dort zahlreiche exotische Pflanzen und Tiere aufspüren.

Koh Samet

Koh Samet

A los buceadores les encanta el diverso mundo submarino con sus jardines de coral. Los amantes del sol hacen una peregrinación a Koh Samet para relajarse en las playas de arena bordeadas de árboles de niaouli, siempre verdes. La isla fascina por sus encantadoras bahías y una buena oferta de deportes acuáticos. En la capital Na Dan casi todos los ferris llegan desde el continente. La mayor parte del bullicio prevalece aquí y la densidad de restaurantes y bares es comparativamente alta. El interior de la isla, perteneciente al parque nacional Khao Laem Ya y su respectiva isla, Mu Ko Samet, está prácticamente intacto. Allí, se pueden encotrar muchas plantas y animales exóticos dando un paseo.

Koh Samet

Di Koh Samet i sub amano il variegato mondo sottomarino con i suoi giardini di corallo, mentre chi preferisce il relax viene qui per le spiagge sabbiose fiancheggiate da alberi di cajeput. Con le sue baie incantevoli e una buona offerta di sport acquatici, l'isola è un luogo piacevole da visitare. Nella città di Na Dan fanno scalo quasi tutti i traghetti che provengono dalla terraferma. Qui la vita è molto frenetica e la densità di ristoranti e bar è relativamente alta. L'interno dell'isola, su cui si estende il Parco nazionale di Khao Laem Ya - Mu Ko Samet, è invece poco frequentato e si presenta in gran parte incontaminato. Camminando è possibile vedere molte piante e animali esotici.

Koh Samet

Duikers houden van de gevarieerde onderwaterwereld met zijn koraaltuinen en zonaanbidders trekken massaal naar Koh Samet om te ontspannen op de zandstranden met groenblijvende cajeputbomen. Het eiland betovert met zijn charmante baaien en een goed aanbod aan watersportmogelijkheden. In de hoofdplaats Na Dan komen bijna alle veerboten vanaf het vasteland aan. Hier heerst de grootste drukte en de dichtheid aan restaurants en bars is relatief hoog. Het binnenland van het eiland, dat deel uitmaakt van het nationale park Khao Laem Ya – Mu Ko Samet, is daarentegen grotendeels onaangetast. Als u te voet gaat, kunt u veel exotische planten en dieren aantreffen.

Koh Samet

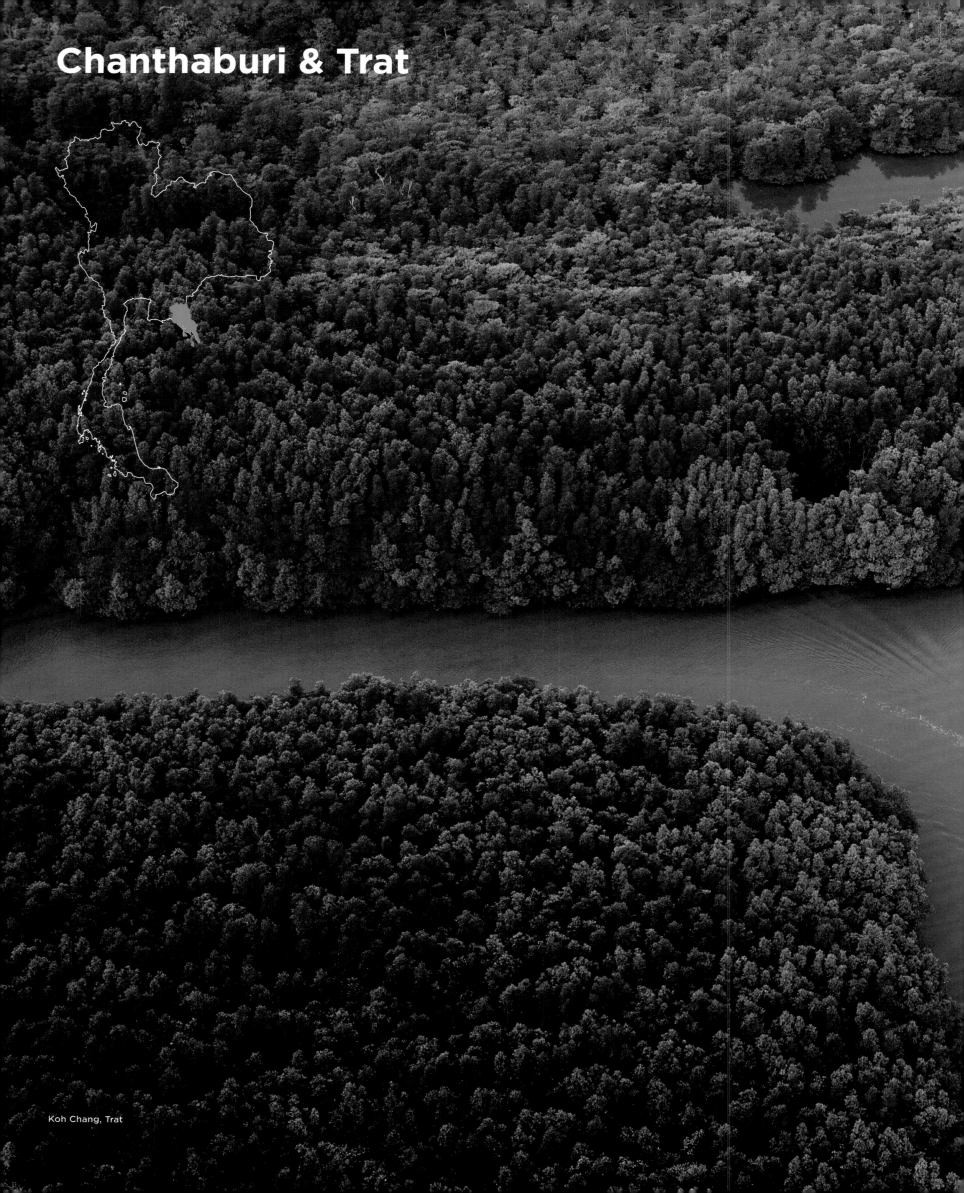

Chanthaburi & Trat

Koh Chang, Trat

Chantaburi

Chanthaburi & Trat

With its gently rolling hills and attractive town of Chanthaburi, the province of the same name welcomes the morning and evening. In the cosy old town, the largest cathedral in Thailand, built in 1880, shines alongside the typical Thai golden chedis. The province of Trat, further south on the border with Cambodia, has dozens of islands. Among them are Koh Chang, Thailand's second largest island, and the much smaller Koh Wai. A tropical paradise with shallow lagoons, the finest sandy beaches and idyllic accommodation. Perfect for snorkelling, diving and relaxing.

Chanthaburi & Trat

Avec ses collines doucement vallonnées et la jolie ville de Chanthaburi, la province du même nom est accueillante le matin comme le soir. Dans l'agréable vieille ville, la plus grande cathédrale de Thaïlande, construite en 1880, brille aux côtés du chedi doré typiquement local. La province de Trat, plus au sud, à la frontière avec le Cambodge, compte des dizaines d'îles, parmi lesquelles Koh Chang, la deuxième plus grande île de Thaïlande, ou la beaucoup plus petite Koh Wai. Cette dernière est un véritable paradis tropical, avec des lagons peu profonds, de magnifiques plages de sable fin et des hôtels de rêve. Parfait pour le snorkeling, la plongée et la détente.

Chanthaburi & Trat

Mit leicht geschwungener Hügellandschaft und der anziehenden Stadt Chanthaburi begrüßt die gleichnamige Provinz den Morgen und den Abend. In der behaglichen Altstadt glänzt neben den typisch thailändischen goldenen Chedis die größte Kathedrale Thailands, die 1880 erbaut wurde. Die weiter südlich gelegene Provinz Trat, an der Grenze zu Kambodscha, hält dutzende Inseln parat. Darunter finden sich Koh Chang, die zweitgrößte Insel Thailands und die wesentlich kleinere Koh Wai. Ein tropisches Paradies mit seichten Lagunen, feinstem Sandstrand und idyllischen Unterkünften. Perfekt geeignet zum Schnorcheln, Tauchen und Erholen.

Koh Wai, Trat

Chanthaburi & Trat

Con sus suaves colinas y la atractiva ciudad de
Chanthaburi, la provincia del mismo nombre da
la bienvenida mañana y tarde. En el acogedor
casco antiguo, la catedral más grande de Tailandia,
construida en 1880, brilla junto a los típicos chedis
dorados tailandeses. La provincia de Trat, más al sur,
en la frontera con Camboya, tiene docenas de islas.
Entre ellas, se encuentra Koh Chang, la segunda isla
más grande de Tailandia y otra mucho más pequeña
llamada Koh Wai. Esta es un paraíso tropical con
lagunas poco profundas, las mejores playas de
arena e idílicos alojamientos. Perfecto para practicar
snorkel, bucear y relajarse.

Chanthaburi & Trat

Con le sue dolci colline e l'attraente città di
Chanthaburi, la provincia omonima ha molto da
offrire. Nell'accogliente centro storico, la più grande
cattedrale della Thailandia, costruita nel 1880, si
erge accanto al tipico chedi d'oro thailandese.
La provincia di Trat, più a sud al confine con la
Cambogia, è costituita da decine di isole, pronte ad
accogliere i visitatori. Tra queste ci sono Koh Chang,
la seconda isola più grande della Thailandia, e Koh
Wai, decisamente più piccola, veri e propri paradisi
tropicali con tanto di lagune poco profonde, spiagge
di sabbia fine e alloggi idilliaci. Il posto è perfetto per
lo snorkeling, le immersioni e il relax.

Chanthaburi & Trat

Met zijn zacht glooiende heuvels en de aantrekkelijke
stad Chanthaburi verwelkomt de gelijknamige
provincie de ochtend en avond. In de gezellige oude
stad schittert de grootste kathedraal van Thailand,
die in 1880 werd gebouwd, naast de typisch Thaise
gouden chedi's. De verder naar het zuiden gelegen
provincie Trat, aan de grens met Cambodja, beschikt
over tientallen eilanden. Daartoe behoren Koh
Chang, het op een na grootste eiland van Thailand,
en het veel kleinere Koh Wai. Een tropisch paradijs
met ondiepe lagunes, de mooiste zandstranden en
idyllische accommodaties. Perfect om te snorkelen,
duiken en relaxen.

Aquaculture fish and oyster farming, Chantaburi

Chantaboon Old Town, Chantaburi

Noen Nang Phaya Viewpoint, Chantaburi

Koh Mak and the Gulf of Thailand

Directly on the coastal road shortly before the city Chantaburi, you can reach the well known as Noen Nang Phaya lookout. A stop is a must, even if you will rarely enjoy the charming view of the Gulf of Thailand on your own. About 130 km further south, off the coast of Trat, the tiny island of Koh Mak rises from the sea. Visitors can expect peace and a peaceful atmosphere. Hectic is a foreign word. This tiny patch of land with countless coconut palms offers impressive natural beaches and water of the finest quality as a happy addition.

Koh Mak et le Golfe de Thaïlande

Le belvédère Noen Nang Phaya se trouve directement sur la route côtière, juste avant la ville de Chantaburi. L'arrêt est un must, même si vous serez rarement seuls à apprécier la vue charmante sur le golfe de Thaïlande. Environ 130 km plus au sud, au large de Trat, se trouve la petite île de Koh Mak, qui propose à ses visiteurs une atmosphère calme et paisible : l'agitation y est un mot inconnu. Ce petit coin de paradis aux innombrables cocotiers offre des plages naturelles impressionnantes mais aussi une eau limpide.

Koh Mak und der Golf von Thailand

Direkt an der Küstenstraße kurz vor der Stadt Chantaburi erreicht man den gut besuchten Noen Nang Phaya Aussichtspunkt. Anhalten ist ein Muss, auch wenn man den reizvollen Blick auf den Golf von Thailand selten für sich allein genießen wird. Gut 130 km weiter südlich erhebt sich vor der Küste von Trat die winzige Insel Koh Mak aus dem Meer. Die Besucher erwartet Ruhe und eine friedvolle Stimmung. Hektik ist ein Fremdwort. Das mit unzähligen Kokospalmen bestückte Fleckchen Erde bietet als gelungene Zugabe eindrucksvolle Naturstrände und Wasserqualität vom Feinsten.

Ao Katueng Beach, Koh Mak, Trat

Koh Mak y el Golfo de Tailandia

Directamente por la carretera costera, poco antes de la ciudad de Chantaburi se llega al concurrido mirador de Noen Nang Phaya. Esta parada es obligatoria, aunque no se suela disfrutar de la encantadora vista del Golfo de Tailandia de manera individual. Unos 130 km más al sur, frente a la costa de Trat, la diminuta isla de Koh Mak se eleva desde el mar. Los visitantes pueden esperar paz y una atmósfera pacífica. La palabra agitación resulta aquí desconocida. Este pequeño trozo de tierra, con innumerables cocoteros, ofrece impresionantes playas naturales y una calidad inmejorable del agua.

Koh Mak e il golfo di Thailandia

Direttamente sulla strada costiera poco prima della città di Chantaburi si raggiunge il punto di vista Noen Nang Phaya, molto frequentato. Una sosta qui è d'obbligo, anche se raramente si può godere indisturbati dell'incantevole vista sul golfo di Thailandia. Circa 130 km più a sud, al largo della costa di Trat, si trova la piccola isola di Koh Mak. I visitatori vi trovano un'atmosfera calma e distesa – la frenesia non è qui di casa. Questo piccolo angolo di terra con innumerevoli palme da cocco inoltre incanta con le sue splendide spiagge naturali e l'ottima qualità delle acque.

Koh Mak en de Golf van Thailand

Direct aan de kustweg vlak voor de stad Chantaburi bereikt u het goed bezochte uitkijkpunt Noen Nang Phaya. Een oponthoud hier is een must, zelfs als u zelden helemaal alleen kunt genieten van het schitterende uitzicht op de Golf van Thailand. Ongeveer 130 km verder naar het zuiden, voor de kust van Trat, rijst het kleine eiland Koh Mak op uit de zee. Bezoekers kunnen hier rust en een vredige sfeer verwachten. Hectiek is hier een onbekend woord. Dit kleine stukje land met talloze kokospalmen biedt als geslaagde toegift indrukwekkende natuurstranden en water van de beste kwaliteit.

Kung Krabaeng, Chantaburi

Koh Kut, Trat

Koh Kut

Koh Kut is part of the Mu Koh Chang Sea National Park and is located off the coast of Cambodia in the province of Trat. Numerous coconut and rubber plantations are spread over the more than 100 km² island. These products are the main source of income for the inhabitants. The few tourists enjoy the beautiful beaches of the west coast or explore the attractive and lonely bays by kayak. In the barely developed interior of the island, the main attractions are the Khlong Chao and Klong Yaiki waterfalls.

Koh Kut

Koh Kut appartient au parc national marin de Mu Koh Chang ; il est situé au large du Cambodge, dans la province de Trat. De nombreuses plantations de noix de coco et de caoutchouc sont réparties sur la grande île de plus de 100 km², et ces produits sont la principale source de revenus des habitants. Les quelques touristes profitent des belles plages de la côte ouest ou explorent les baies attrayantes et solitaires en kayak. Dans l'intérieur à peine développé de l'île, les principales attractions sont les chutes d'eau de Khlong Chao et Klong Yaiki.

Koh Kut

Koh Kut gehört zum Mu-Koh-Chang-Meeresnationalpark und liegt vor der Küste Kambodschas in der Provinz Trat. Auf der über 100 km² großen Insel verteilen sich zahlreiche Kokos- und Kautschukplantagen. Diese Produkte sind die Haupterwerbsquelle der Einwohner. Die wenigen Touristen erfreuen sich an den herrlichen Stränden der Westküste oder erkunden die anziehenden und einsamen Buchten mit dem Kajak. Im kaum erschlossenen Inselinneren sind die Hauptattraktionen der Khlong Chao- und der Klong Yaiki-Wasserfall.

Koh Kut, Trat

Koh Kut

Koh Koodt (también llamada Koh Kut) pertenece al parque nacional marino de Mu Koh Chang y está situado frente a la costa de Camboya, en la provincia de Trat. Numerosas plantaciones de coco y caucho se extienden a lo largo de más de 100 km² de la isla. Estos productos son la principal fuente de ingresos para los habitantes. Los pocos turistas disfrutan de las hermosas playas de la costa oeste o exploran las atractivas y solitarias bahías en kayak. En el interior de la isla, escasamente desarrollado, las principales atracciones son las cataratas Nam Tok Khlong Chao y Nam Tok Klong Yaiki.

Koh Kut

Koh Kut fa parte del Parco nazionale marino di Mu Koh Chang e si trova al largo della costa della Cambogia, nella provincia di Trat. Sull'isola, di oltre 100 km² di superficie, si estendono numerose piantagioni di cocco e caucciù, la principale fonte di reddito per gli abitanti del posto. I pochi turisti si godono le belle spiagge della costa occidentale o esplorano baie solitarie in kayak. Nell'interno dell'isola, poco sviluppato, le attrazioni principali sono le cascate di Khlong Chao e di Klong Yaiki.

Koh Kut

Koh Kut maakt deel uit van het beschermde zeegebied Mu Koh Chang en ligt voor de kust van Cambodja in de provincie Trat. Talrijke kokos- en rubberplantages zijn verspreid over het meer dan 100 km² grote eiland. Deze producten vormen de belangrijkste bron van inkomsten voor de bewoners. De weinige toeristen genieten van de prachtige stranden aan de westkust of verkennen de aantrekkelijke en eenzame baaien in een kajak. In het nauwelijks ontwikkelde binnenland zijn de belangrijkste bezienswaardigheden de watervallen Khlong Chao en Klong Yaiki.

Koh Kut, Trat

Koh Mak, Trat

Small Island naer Koh Rang, Trat

Koh Mak and Koh Rank

Off the coast of Trat, inhabited and uninhabited island pearls adorn the Gulf of Thailand. With an area of 16 km², Koh Mak, also known as Betel Nut Island, is a manageable paradise. The neighbouring island Koh Rang fascinates divers mainly because of its tropical shoals of fish.

Koh Mak y Koh Rank

Frente a la costa de Trat, las perlas de las islas habitadas y deshabitadas adornan el Golfo de Tailandia. Koh Mak, también llamada la isla de la nuez de areca, es con un área de 16 km² un paraíso manejable. La isla vecina Koh Rang fascina a los buceadores principalmente por sus bancos tropicales de peces.

Koh Mak et Koh Rank

Au large de la côte de Trat, des joyaux déserts et inhabités illuminent le golfe de Thaïlande. Koh Mak, également appelée « l'île aux noix de Betel », est, avec une superficie de 16 km², un paradis à taille humaine. L'île voisine Koh Rang fascine les plongeurs principalement grâce à ses bancs de poissons tropicaux.

Koh Mak e Koh Rank

Al largo della costa di Trat, isole affascinanti, abitate e non, adornano il golfo di Thailandia. Koh Mak, detta anche "isola della palma di Betel", è un piccolo paradiso, con una superficie di soli 16 km². La vicina isola di Koh Rang attrae i sub soprattutto per i suoi banchi di pesci tropicali.

Koh Mak und Koh Rang

Vor der Küste von Trat schmücken bewohnte und unbewohnte Inselperlen den Golf von Thailand. Koh Mak, auch Betelnussinsel genannt, ist mit einer Fläche von 16 km² ein überschaubares Paradies. Die Nachbarinsel Koh Rang zieht vor allem wegen ihrer tropischen Fischschwärme Taucher in ihren Bann.

Koh Mak en Koh Rank

Voor de kust van Trat sieren bewoonde en onbewoonde eilanden de Golf van Thailand. Koh Mak, ook wel Betelnooteiland genoemd, is met een oppervlakte van 16 km² een overzichtelijk paradijs. Het naburige eiland Koh Rang fascineert duikers vooral vanwege de tropische scholen vissen.

Ratchaburi & Phetchaburi

Rice field, Phetchaburi

Blue Pitta, Kaeng Krachan National Park, Phetchaburi

Ratchaburi & Phetchaburi

Both of these western provinces are on the border with Myanmar's mountainous natural areas, but also have extensive plains crossed by rivers. The climate is tropical and humid. People make their living from the cultivation of rice, sugar cane, fruit and tourism. The limestone landscape conceals a wide variety of cave systems, and majestic temples enthrone the hills. Nature lovers spend their time in the national parks watching birds or looking for bears or tigers.

Ratchaburi & Phetchaburi

Les deux provinces occidentales sont situées à la frontière avec les montagnes du Myanmar, mais elles contiennent aussi de vastes plaines traversées par des rivières. Le climat est tropical et humide. Les habitants vivent de la culture du riz, de la canne à sucre, des fruits et du tourisme. Dans ce paysage en partie calcaire, il est possible d'admirer des systèmes de grottes très ramifiés mais aussi des temples majestueux qui trônent sur les collines. Les amoureux de la nature pourront passer du temps dans les parcs nationaux à observer les oiseaux ou à partir à la recherche d'ours et de tigres.

Ratchaburi & Phetchaburi

Beide Westprovinzen sind an der Grenze zu Myanmars bergiger Natur, verfügen aber auch über weitläufige, von Flüssen durchzogene Ebenen. Das Klima ist tropisch-feucht. Die Menschen leben vom Anbau von Reis, Zuckerrohr, Obst und vom Tourismus. In der teils von Kalkstein geprägten Landschaft verstecken sich weitverzweigte Höhlensysteme und auf den Hügeln thronen majestätische Tempelanlagen. Naturfreunde verbringen die Zeit in den Nationalparks und gehen auf Vogelbeobachtung oder halten Ausschau nach Bären oder Tigern.

Phra Nakhon Khiri Historical Park, Phetchaburi

Ratchaburi & Phetchaburi

Ambas provincias occidentales están en la frontera con la naturaleza montañosa de Myanmar, pero también tienen extensas llanuras atravesadas por ríos. El clima es húmedo- tropical. La gente vive del cultivo del arroz, la caña de azúcar, la fruta y el turismo. En el paisaje, caracterizado en parte por la caliza, hay sistemas de cuevas muy ramificadas y majestuosos templos encumbrando las colinas. Los amantes de la naturaleza pasan su tiempo en los parques nacionales observando aves o buscando osos o tigres.

Ratchaburi & Phetchaburi

Entrambe le province occidentali confinano con il Myanmar, a cui sono accomunate dalla regione montuosa. Non mancano però anche vaste pianure attraversate da fiumi. Il clima qui è tropicale-umido. La gente vive di turismo e della coltivazione di riso, canna da zucchero e frutta. Il paesaggio, in parte calcareo, è caratterizzato da sistemi di grotte molto ramificati e da maestosi templi che sovrastano dall'alto delle colline. Gli amanti della natura hanno la possibilità di visitare diversi parchi nazionali ,dove osservare uccelli, orsi e tigri.

Ratchaburi & Phetchaburi

Beide westelijke provincies liggen op de grens met het bergachtige landschap van Myanmar, maar beschikken ook over uitgestrekte vlaktes die worden doorkruist door rivieren. Het klimaat is tropisch vochtig. Mensen leven van de teelt van rijst, suikerriet, fruit en toerisme. In het landschap, dat deels wordt gekenmerkt door kalksteen, bevinden zich wijdvertakte grotsystemen en tronen majestueuze tempels op de heuvels. Natuurliefhebbers brengen hun tijd door in de nationale parken, waar ze naar vogels kijken of op zoek gaan naar beren of tijgers.

Floating Market, Damnoen Saduak, Ratchaburi

Floating Market, Damnoen Saduak, Ratchaburi

Floating Market

Known and unknown vegetable and fruit varieties
in all colours and shapes, and in between kitchens
housed on swaying boats. A feast for all the senses.
Exotic smells and some flavours that may take some
getting used to await the visitor at the charming and
bustling floating market of Damnoen Saduak.

Mercado Flotante

Variedades conocidas y desconocidas de verduras
y frutas de todos los colores y formas. Todo ello,
formando bodegones a bordo de barcos inestables.
Un festín para los sentidos. Olores exóticos y algunos
matices gustativos que acostumbran esperan al
visitante en el fascinante y bullicioso mercado flotante
de Damnoen Saduak.

Marché flottant

Des variétés connues et inconnues de légumes et
de fruits, de toutes les couleurs et de toutes les
formes et, au milieu, sur des bateaux, des stands de
nourriture : les marchés flottants sont un festin pour
tous les sens. Des odeurs exotiques et des goûts
nouveaux attendent ainsi le visiteur au charmant et
animé marché flottant de Damnoen Saduak.

Mercato galleggiante

Varietà di frutta e verdura più o meno conosciute, di
qualsiasi colore e forma, cucine improvvisate a bordo
di barche ondeggianti: ricco di profumi esotici e
odori che però non sempre risultano subito piacevoli,
quella che accoglie il visitatore al vivace mercato
galleggiante di Damnoen Saduak è una vera e propria
festa dei sensi.

Schwimmender Markt

Bekannte und unbekannte Gemüse- und Obstsorten
in allen Farben und Formen. Dazwischen auf
schwankenden Booten untergebrachte Garküchen.
Ein Fest für alle Sinne. Exotische Gerüche und manch
gewöhnungsbedürftige Geschmacksnuance erwarten
den Besucher auf dem charmanten und betriebsamen
schwimmenden Markt von Damnoen Saduak.

Drijvende Markt

Bekende en onbekende groente- en fruitsoorten in
alle kleuren en vormen. Daartussen eetkraampjes die
zijn ondergebracht op wiebelende bootjes. Een feest
voor alle zintuigen. De bezoekers van de charmante
en bruisende drijvende markt van Damnoen Saduak
kunnen exotische geuren verwachten en een aantal
smaaknuances waar ze wel aan moeten wennen.

Khao Ngu Stone Park, Ratchaburi

Bangtaboon Bay, Phetchaburi

The coast of Phetchaburi

In addition to fishing, some families in the province of Phetchaburi run large salt farms to earn a living. The farmers direct sea water into so-called salt gardens, where it slowly evaporates. What remains is the white gold, which is laboriously piled up into hundreds of mounds.

La côte de Phetchaburi

En plus de la pêche, certaines familles de la province de Phetchaburi exploitent pour gagner leur vie de grandes fermes salines. Les agriculteurs conduisent l'eau de mer vers ce qu'on appelle les jardins de sel, où elle s'évapore lentement. Reste alors « l'or blanc », laborieusement empilé en centaines de monticules.

Die Küste von Phetchaburi

Neben dem Fischfang betreiben einige Familien in der Provinz Phetchaburi große Salzfarmen, um ihren Lebensunterhalt zu bestreiten. Die Bauern leiten das Meerwasser in sogenannte Salzgärten, wo es langsam verdunstet. Zurück bleibt das weiße Gold, welches in mühevoller Handarbeit auf hunderte Haufen aufgetürmt wird.

Sea Salt Farming, Phetchaburi

La costa de Phetchaburi

Además de la pesca, algunas familias de la provincia de Phetchaburi tienen grandes granjas de sal para ganarse la vida. Los agricultores llevan el agua de mar a los llamados jardines de sal, donde se evapora lentamente. Lo que queda es el oro blanco, que está laboriosamente amontonado en cientos de pilas.

La costa di Phetchaburi

Oltre a dedicarsi alla pesca, alcune famiglie nella provincia di Phetchaburi gestiscono grandi saline per guadagnarsi da vivere. I contadini hanno incanalato l'acqua del mare verso i cosiddetti giardini di sale, dove lentamente evapora. Rimane l'oro bianco, che faticosamente viene riunito in piccoli mucchi.

De kust van Phetchaburi

Naast de visserij bestieren sommige families in de provincie Phetchaburi grote zoutboerderijen om in hun levensonderhoud te voorzien. De boeren leiden het zeewater naar zogenaamde zouttuinen, waar het langzaam verdampt. Wat overblijft, is het witte goud, waarvan met de hand moeizaam honderden hopen worden gemaakt.

Kaeng Krachan Dam, Kaeng Krachan National Park, Phetchaburi

Oriental pied hornbill, Kaeng Krachan National Park, Phetchaburi

Kaeng Krachan National Park

Kaeng Krachan is Thailand's largest national park, with an area of almost 3000 km². In the mountains, which are partly covered by rainforest and rise to an altitude of 1500 m on the border to Myanmar, live hundreds of bird species such as the Oriental Hornbill. Its beak reaches a length of up to 20 cm. Another inhabitant of the reserve is the Malay porcupine. When in danger, it raises its spines and makes grunting noises. The rodents, which mainly go foraging at night, reach a height of up to 75 cm and live with their families.

Parc national de Kaeng Krachan

Kaeng Krachan est, avec une superficie de près de 3000 km², le plus grand parc national de Thaïlande. Des centaines d'espèces d'oiseaux vivent dans ses montagnes à la frontière du Myanmar, hautes de 1500 m et en partie recouvertes de forêt tropicale humide. On y rencontre ainsi des calaos orientaux, dont les becs peuvent atteindre 20 cm. Parmi les résidents de la réserve, on trouve également le porc-épic malais. Lorsqu'il est en danger, il érige ses pointes et pousse des grognements. Les rongeurs, qui se nourrissent principalement la nuit, atteignent une hauteur de 75 cm et vivent avec leur famille.

Kaeng Krachan-Nationalpark

Der Kaeng Krachan ist mit einer Fläche von knapp 3000 km² der größte Nationalpark Thailands. In den teils von Regenwald bedeckten Bergen, die an der Grenze zu Myanmar bis auf eine Höhe von 1500 m aufragen, leben hunderte Vogelarten wie der oft anzutreffende Orienthornvogel. Sein Schnabel erreicht eine Länge von bis zu 20 cm. Ein anderer Bewohner des Schutzgebietes ist das Malaiische Stachelschwein. Bei Gefahr stellt es eine Stacheln auf und gibt grunzende Geräusche von sich. Die hauptsächlich nachts auf Nahrungssuche gehenden Nagetiere erreichen eine Körpergröße von bis zu 75 cm und leben im Familienverbund.

Malayan or Himalayan porcupine, Kaeng Krachan National Park, Phetchaburi

Parque nacional Kaeng Krachan

Kaeng Krachan es el parque nacional más grande de Tailandia, con una superficie de casi 3000 km². En las montañas, que están cubiertas en parte por la selva tropical y se elevan a una altitud de 1500 m en la frontera con Myanmar, viven centenares de especies de aves, como el cálao cariblanco, al que se puede avistar a menudo. Su pico alcanza una longitud de hasta 20 cm. Otro habitante de la reserva es el puercoespín malayo. Cuando está en peligro, erige las púas y gruñe.Se alimentan principalmente de Los roedores que caza por la noche. Los puercoespines llegan a alcanzar un tamaño dehasta 75 cm y viven con sus familias.

Parco nazionale Kaeng Krachan

Kaeng Krachan è il più grande parco nazionale della Thailandia, con una superficie di quasi 3.000 km². Sui monti in parte ricoperti dalla foresta pluviale al confine con il Myanmar, alti fino a 1.500 m, vivono centinaia di specie di uccelli come il molto diffuso bucero bicorne, il cui becco raggiunge una lunghezza di 20 cm. Un altro abitante della riserva è l'istrice malese: quando è in pericolo, il roditore innalza i suoi aculei ed emette un verso. Questi animali, che di notte si dedicano principalmente alla ricerca di cibo, raggiungono un'altezza di 75 cm e vivono in famiglia.

Nationaal park Kaeng Krachan

Kaeng Krachan is met een oppervlakte van bijna 3000 km² het grootste nationale park van Thailand. In de bergen, die gedeeltelijk bedekt zijn met regenwoud en tot 1500 meter hoogte oprijzen aan de grens met Myanmar, leven honderden vogelsoorten, zoals de bonte neushoornvogel, die hier veel wordt aangetroffen. Zijn snavel kan 20 cm lang worden. Een andere bewoner van het natuurreservaat is het Maleis stekelvarken. Bij gevaar gaan zijn stekels rechtop staan en brengt het dier een knorrend geluid voort. Deze knaagdieren, die voornamelijk 's nachts op zoek gaan naar eten, bereiken een lichaamshoogte van 75 cm en leven in familieverband.

Bryes whales

Kaeng Krachan Dam, Kaeng Krachan National Park, Phetchaburi

Kaeng Krachan Dam

Just outside Kaeng Krachan National Park the dam
of the same name, built in 1966, is used to generate
electricity. It is fed by the Phetchaburi River, which
rises from the Tanaosri Mountains. It reaches as far
as Myanmar and later flows into the Gulf of Thailand.
There are some small islands in the reservoir.
Visitors can engage in activities on the water, and a
lookout point on the dam provides a great view over
the landscape.

Barrage de Kaeng Krachan

Juste à l'extérieur du parc national de Kaeng Krachan,
le barrage du même nom a été installé en 1966 pour
produire de l'électricité. Il est alimenté par le fleuve
Phetchaburi, qui prend sa source dans les montagnes
Tenasserim. Il atteint le Myanmar et se jette ensuite
dans le golfe de Thaïlande. Il y a quelques petites îles
dans le réservoir. Les visiteurs peuvent participer à
des activités nautiques et un belvédère construit sur le
barrage offre une vue imprenable sur le paysage.

Kaeng Krachan-Staudamm

Etwas außerhalb des Kaeng Krachan-Nationalparks
breitet sich der gleichnamige, 1966 erbaute
Staudamm aus, der der Erzeugung von Strom dient.
Gespeist wird er vom Phetchaburi-Fluss, der dem
bis nach Myanmar reichenden Tanaosri-Gebirge
entspringt und im späteren Verlauf in den Golf von
Thailand mündet. Im Staubecken befinden sich
einige kleine Inseln. Besucher können Aktivitäten auf
dem Wasser unternehmen und ein auf dem Damm
errichteter Aussichtspunkt sorgt für einen tollen Blick
über die Landschaft.

Kaeng Krachan Dam, Kaeng Krachan National Park, Phetchaburi

Presa de Kaeng Krachan

A las afueras del parque nacional Kaeng Krachan se sitúa la presa del mismo nombre, construida en 1966, se creó para generar electricidad. La alimenta el río Phetchaburi, que nace en las montañas de Tanaosri y llega hasta Myanmar para luego desembocar en el golfo de Tailandia, que alberga algunas islas pequeñas. Los visitantes pueden realizar actividades sobre el agua y un mirador construido sobre la presa ofrece una gran vista sobre el paisaje.

Diga di Kaeng Krachan

Appena fuori dal Parco nazionale di Kaeng Krachan si trova l'omonima diga, costruita nel 1966 per generare elettricità. Ad alimentarla è il fiume Phetchaburi, che sgorga dai monti Tanaosri, lungo il confine con il Myanmar, e poi sfocia nel golfo di Thailandia. Il bacino idrico è costellato da alcune piccole isole. I visitatori possono fare sport acquatici, mentre un belvedere costruito sulla diga offre una splendida vista sul paesaggio.

Kaeng Krachan-stuwdam

Net buiten het nationale park Kaeng Krachan staat de gelijknamige stuwdam, die in 1966 is gebouwd om elektriciteit op te wekken. Het stuwmeer krijgt zijn water van de Phetchaburi-rivier, die uit het tot Myanmar reikende Tenasserim-gebergte ontspringt en later uitmondt in de Golf van Thailand. Er zijn enkele kleine eilandjes in het stuwmeer. Bezoekers kunnen watersporten beoefenen op het water, en een uitkijkpunt op de dam biedt een prachtig uitzicht over het landschap.

Tham Khao Luang Cave, Phetchaburi

Tham Khao Luang Cave

Not far from the city of Phetchaburi, a very special spectacle takes place. The stage for this is Tham Khao Luang Cave. As soon as the sun is high in the sky, the sunlight falls atmospherically on the Buddha figures placed inside and lights them up in a special atmosphere. An extraordinary place for contemplation and peace.

Cueva Tham Khao Luang

No lejos de la ciudad de Phetchaburi tiene lugar un espectáculo muy especial enla Cueva Tham Khao Luang. Tan pronto como el sol está alto en el cielo, la luz del día brilla sobre las figuras de Buda colocadas en su interior. Un lugar extraordinario de contemplación y paz.

Grotte de Tham Khao Luang

Non loin de la ville de Phetchaburi, la grotte de Tham Khao Luang abrite un spectacle très spécial : lorsque le soleil est haut dans le ciel, la lumière du jour rayonne depuis une ouverture au plafond sur les statues de Bouddha placées à l'intérieur. Un lieu extraordinaire de contemplation et de paix.

Grotte di Tham Khao Luang

Non lontano dalla città di Phetchaburi è possibile ammirare uno spettacolo molto speciale, a cui fanno da sipario le grotte di Tham Khao Luang: non appena il sole è alto in cielo, la luce risplende sulle statue del Buddha poste all'interno della grotta creando un'atmosfera molto particolare. Un insolito luogo di contemplazione e di pace.

Tham Khao Luang-Höhle

Unweit der Stadt Phetchaburi vollzieht sich ein ganz besonderes Schauspiel. Das Podium dafür bietet die Tham Khao Luang-Höhle. Sobald die Sonne hoch am Himmel steht, fällt das Tageslicht auf die im Inneren aufgestellten Buddhafiguren und erleuchtet sie atmosphärisch. Ein außerordentlicher Ort der Besinnung und Ruhe.

Tham Khao Luang-grot

Niet ver van de stad Phetchaburi vindt een heel bijzonder spektakel plaats. De Tham Khao Luang-grot biedt daarvoor het podium. Zodra de zon hoog aan de hemel staat, valt het daglicht op de Boeddhabeelden in de grot en verlicht die heel sfeervol. Een buitengewone plek van bezinning en rust.

WAT YANNAWA

WAT SRISUPHAN

WAT BAN RAI

WAT PHRA THAT DOI KONG MU

WAT CHIANG MAN

WAT PHRA THAT DOI KHAM

WAT PLAI LAEM

WAT PHRA YAI KO PAN

PHRA MAHA CHEDI CHAI MONGKOL

WAT PHRA DHAMMAKAYA

WAT YANSANGWARARAM

WAT TRAIMIT

WAT SAKET

WAT BENCHAMABOPHIT

WAT PHRA YAI

TEMPLES

Thailand's temples are as diverse as the country's regions. Many are centuries old and are surrounded by a magical atmosphere. Others are only a few years old and have a ufo-like shape. Golden Buddha figures, mystical beings and richly decorated, conically shaped chedis characterize almost every one of these religious sanctuaries.

TEMPLES

Il existe autant de sortes de temples que de régions dans le pays. Beaucoup sont centenaires et auréolés d'une atmosphère magique. D'autres n'ont que quelques années et ont parfois des allures d'ovni. Dans quasiment tous ces sanctuaires religieux, on trouve des figures des Bouddhas d'or, des statues d'êtres mystiques et des chedis de forme conique richement décorés.

TEMPEL

Die Tempel Thailands sind so vielfältig wie die Regionen des Landes. Viele sind Jahrhunderte alt und von magischer Atmosphäre umgeben. Andere wenige Jahre jung und von ufo-ähnlicher Gestalt. Goldene Buddhafiguren, mystische Wesen und reich verzierte, konisch geformte Chedis prägen nahezu jedes dieser religiösen Heiligtümer.

TEMPLOS

Los templos de Tailandia son tan diversos como las regiones del país. Muchos tienen siglos de antigüedad y están rodeados de una atmósfera mágica. Otros son modernos y tienen forma de OVNI. Las figuras doradas de Buda, los seres místicos y los chedis de forma cónica muy ornamentados caracterizan casi todos estos santuarios religiosos.

TEMPLI

I templi della Thailandia sono variegati come le regioni del paese. Circondati da un'atmosfera magica, molti esistono da secoli, altri sono solo di poco più giovani e simili a ufo. Buddha dorati, figure mistiche e chedi a forma di cono riccamente decorati caratterizzano quasi tutti i santuari spirituali.

TEMPELS

De tempels van Thailand zijn net zo verschillend als de regio's van het land. Veel ervan zijn eeuwen oud en omgeven door een magische sfeer. Andere zijn nog maar een paar jaar oud en hebben een ufo-achtige vorm. Gouden Boeddhabeelden, mystieke wezens en rijkversierde, kegelvormige chedi's kenmerken bijna al deze religieuze heiligdommen.

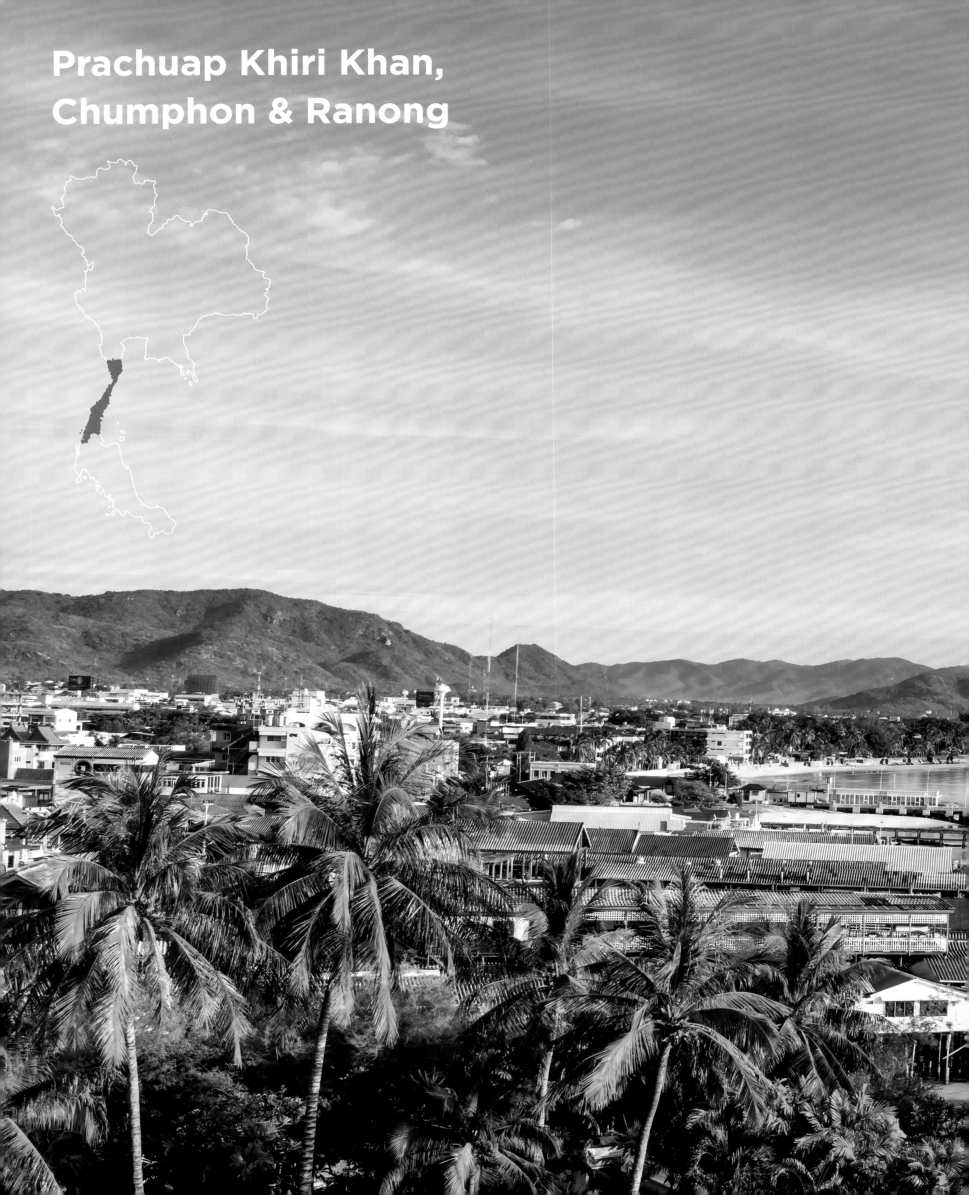

Prachuap Khiri Khan, Chumphon & Ranong

Wat Thammikaram Worawihan, Prachuap Khiri Khan

Phraya Nakhon Cave, Khao Sam Roi Yot National Park,
Prachuap Khiri Khan

Prachuap Khiri Khan, Chumphon & Ranong
Hua Hin, Thailand's oldest seaside resort, is one of the most beautiful cities in the province of Prachuap Khiri Khan, whose Khao Sam Roi Yot National Park is also worth a visit. Chumphon and Ranong offer great diving sites and a unique island world, in both the eastern Gulf of Thailand and the western Andaman Sea.

Prachuap Khiri Khan, Chumphon & Ranong
Hua Hin, plus ancienne station balnéaire de Thaïlande, est l'une des plus belles villes de la province de Prachuap Khiri Khan, dont le parc national de Khao Sam Roi Yot mérite également une visite. Chumphon et Ranong offrent de magnifiques sites de plongée mais également des paysages insulaires uniques, que ce soit dans l'est du Golfe de Thaïlande ou dans l'ouest de la mer d'Andaman.

Prachuap Khiri Khan, Chumphon & Ranong
Hua Hin, Thailands ältester Badeort, gehört zu den schönsten Städten in der Provinz Prachuap Khiri Khan, deren Khao Sam Roi Yot Nationalpark ebenfalls einen Besuch lohnt. Chumphon und Ranong stehen für grandiose Tauchgründe und eine einzigartige Inselwelt, egal ob im östlich gelegenen Golf von Thailand oder in der sich westlich zeigenden Andamanensee.

Prachuap Khiri Khan, Chumphon & Ranong
Hua Hin, el balneario más antiguo de Tailandia, está en una de las ciudades más bellas de la provincia de Prachuap Khiri Khan, cuyo parque nacional de Khao Sam Roi Yot también merece una visita. Chumphon y Ranong son sinónimos de magníficos lugares para bucear y de un mundo insular único, ya sea al este del golfo de Tailandia o al oeste del mar de Andamán.

Prachuap Khiri Khan, Chumphon & Ranong
Hua Hin, la più antica località balneare della Thailandia, è anche una delle città più belle della provincia di Prachuap Khiri Khan. Qui, anche il Parco nazionale di Khao Sam Roi Yot merita una visita. Chumphon e Ranong sono sinonimo di magnifici luoghi per le immersioni e di un mondo insulare unico, sia sul litorale a est, nel golfo di Thailandia, che su quello a ovest, che si affaccia sul mare delle Andamane.

Prachuap Khiri Khan, Chumphon & Ranong
Hua Hin, de oudste badplaats van Thailand, is een van de mooiste steden in de provincie Prachuap Khiri Khan, waar het nationale park Khao Sam Roi Yot ook zeker een bezoek waard is. Chumphon en Ranong staan voor prachtige duiklocaties en een unieke eilandenwereld, zowel in de oostelijk gelegen Golf van Thailand als in de westelijke Andamanse Zee.

View from Khao Lom Muak Mountain, Prachuap Khiri Khan

Khao Sam Roi Yot National Park, Prachuap Khiri Khan

Khao Sam Roi Yot National Park

Hundreds of limestone hills, some over 600 m high, give Thailand's first maritime national park its name. This protected area on the Gulf of Thailand delights with mangrove forests, fresh water swamps and beautiful beaches. The best way to discover the region is on the established paths or by boat. The bird life is particularly remarkable. More than 300 species live here, including the Malay Plover, the White-bellied Marsh Eagle and the rare Spoon-billed Sandpiper, a migratory bird that spends the summer in Russia.

Parc national de Khao Sam Roi Yot Yot

Des centaines de collines calcaires, certaines de plus de 600 m de haut, ont donné son nom au premier parc national maritime de Thaïlande – il signifie « la montagne au 300 pics ». L'aire protégée sur le golfe de Thaïlande contient des forêts de mangroves, des marécages d'eau douce et de belles plages. Les meilleures façons de découvrir la région sont les sentiers ou le bateau. L'avifaune est particulièrement remarquable : plus de 300 espèces y vivent, dont le pluvier malais, l'aigle des marais à ventre blanc et le rare bécasseau à bec cuillère, un oiseau migrateur qui passe l'été en Russie.

Khao Sam Roi Yot Nationalpark

Hunderte Kalksteinhügel, einige über 600 m hoch, gaben dem ersten maritimen Nationalpark Thailands seinen Namen. Das Schutzgebiet am Golf von Thailand begeistert durch Mangrovenwälder, Süßwassersümpfe und schöne Strände. Die Region lässt sich am besten auf den angelegten Wegen oder bei einer Bootsfahrt entdecken. Bemerkenswert kommt vor allem die Vogelwelt daher. Über 300 Arten leben hier, darunter der Malaiische Regenpfeifer, der Weißbauch-Seeadler und der seltene Löffelschnabelstrandläufer, ein Zugvogel, der den Sommer in Russland verbringt.

Khao Sam Roi Yot National Park, Prachuap Khiri Khan

Parque nacional Khao Sam Roi Yot

Cientos de colinas de piedra caliza, algunas de más de 600 m de altura, dieron nombre al primer parque nacional marítimo de Tailandia. El área protegida en el golfo de Tailandia se deleita con bosques de manglares, pantanos de agua dulce y hermosas playas. La mejor manera de descubrir la región es por los senderos o en barco. La avifauna es particularmente notable. Más de 300 especies viven aquí, incluyendo el chorlito malayo, el águila pescadora de vientre blanco y el raro correlimos de pico cuchara, un ave migratoria que pasa el verano en Rusia.

Parco nazionale Khao Sam Roi Yot

Centinaia di colline calcaree, alcune alte oltre 600 m, hanno dato il nome al primo parco nazionale marittimo della Thailandia. L'area protetta che si estende lungo il golfo di Thailandia è ricca di foreste di mangrovie, paludi d'acqua dolce e splendide spiagge. Il modo migliore per scoprire la regione è lungo i sentieri o in barca. L'avifauna è particolarmente notevole: qui vivono più di 300 specie, tra cui il corriere della Malaysia, l'aquila pescatrice panciabianca e il raro gambecchio becco a spatola, un uccello migratore che trascorre l'estate in Russia.

Nationaal park Khao Sam Roi Yot

Honderden kalksteenheuvels, waarvan enkele meer dan 600 meter hoog, gaven het eerste maritieme nationale park van Thailand zijn naam. Het beschermde zeegebied aan de Golf van Thailand verrukt met zijn mangrovebossen, zoetwatermoerassen en prachtige stranden. De beste manier om de regio te ontdekken is over de aangelegde wegen of per boot. Vooral het vogelleven is opmerkelijk. Hier leven meer dan 300 soorten, waaronder de Maleise plevier, de witbuikzeearend en de zeldzame lepelbekstrandloper, een trekvogel die de zomer doorbrengt in Rusland.

View from Khao Chong Krachok Mountain, Prachuap Khiri Khan

Buddha Statues, Hua Hin, Prachuap Khiri Khan

an Elephants, Kui Buri National Park, Prachuap Khiri Khan

Hin and Kui Buri National Park

a Hin and the surrounding area are home to
ntless temples decorated with beautiful Buddha
tues that are worth seeing, such as Wat Khao
ilas. Kui Buri National Park, which was opened in
9 and is located 80 km south of Hua Hin, is close
nature. During a visit, the chances of seeing one of
more than 300 elephants are extraordinarily good.

que nacional Hua Hin y Kui Buri

a Hin y sus alrededores albergan innumerables
nplos que vale la pena visitar, decorados con
las estatuas de Buda, ejemplo son el templo Wat
ao Krailas. El parque nacional Kui Buri, inaugurado
1999 y situado a 80 km al sur de Hua Hin, está
clavado en la naturaleza. Durante una visita, las
sibilidades de ver uno de los más de 300 elefantes
n extraordinariamente elevadas.

Parc national de Hua Hin et Kui Buri Buri

Hua Hin et ses environs abritent d'innombrables
temples dignes d'être visités et décorés de belles
statues de Bouddha, comme le Wat Khao Krailas.
Le parc national de Kui Buri, ouvert en 1999 et situé
à 80 km au sud de Hua Hin, est assez sauvage et
permet très souvent d'apercevoir l'un des plus de
300 éléphants qui y résident.

Parco nazionale di Hua Hin e Kui Buri

A Hua Hin e dintorni è possibile visitare numerosi
templi, decorati con bellissime statue del Buddha,
come quello di Wat Khao Krailas. Il Parco nazionale di
Kui Buri, che è stato aperto nel 1999 e si trova a 80 km
a sud di Hua Hin, è invece di stampo naturalistico.
Durante la visita, le possibilità di avvistare uno degli
oltre 300 elefanti sono decisamente alte.

Hua Hin und Kui Buri National Park

Hua Hin und die Umgebung beherbergen unzählige
sehenswerte, mit schönen Buddhastatuen
geschmückte Tempel wie den Wat Khao Krailas.
Naturnaher präsentiert sich der waldreiche und 1999
eröffnete Kui Buri Nationalpark, der sich rund 80 km
südlich von Hua Hin befindet. Bei einem Besuch
stehen die Chancen, einen der über 300 Elefanten
zu sehen, außerordentlich gut.

Hua Hin en nationaal park Kui Buri

Hua Hin en de omgeving herbergen talloze
bezienswaardige, met prachtige Boeddhabeelden
versierde tempels, zoals Wat Khao Krailas. Het
bosrijke nationale park Kui Buri, dat in 1999 werd
geopend en 80 km ten zuiden van Hua Hin ligt,
staat dicht bij de natuur. Tijdens een bezoek is de
kans om een van de meer dan 300 olifanten te zien
buitengewoon groot.

Koh Ngam Noi, Chumphon

Hat Bang Ben Beach, Ranong

Ao Khao Kwai (Buffalo Horn Beach), Koh Phayam, Ranong

Khao Fa Chi Viewpoint, Ranong

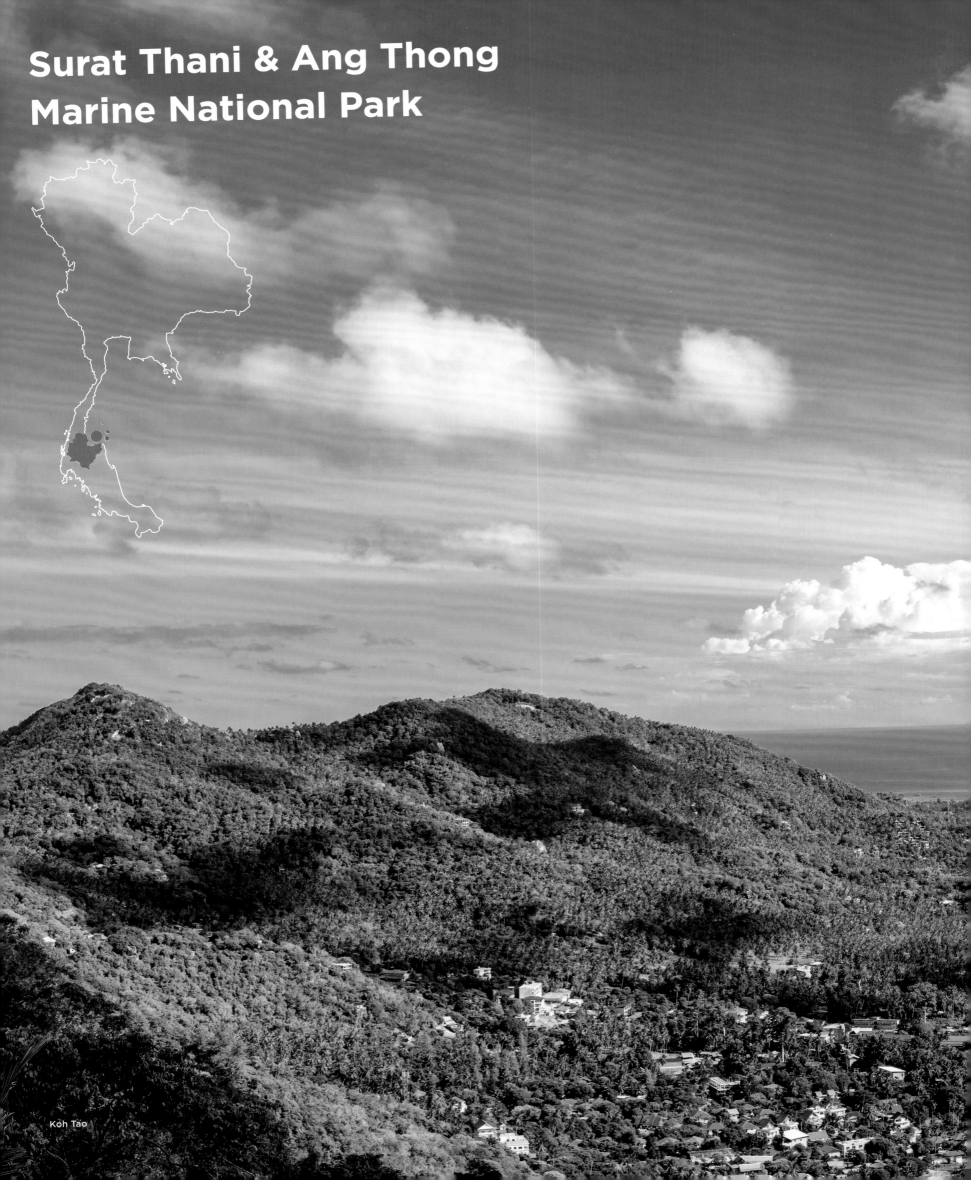

Surat Thani & Ang Thong
Marine National Park

Koh Tao

Surat Thani

Surat Thani & Ang Thong Marine National Park

Surat Thani province includes islands that are like
sand by the sea, including the famous maritime jewels
of Koh Phangan, Koh Tao and Koh Samui, as well as
the dozens of islets of the Ang Thong Marine National
Park. If you want to go on a fishing safari with
your camera, you will hardly find a better region in
Thailand. Khao Sok National Park is a dream in green
with tropical vegetation and secluded waters.

Surat Thani & Ang Thong Marine National Park

La province de Surat Thani comprend des îles qui
ressemblent à des étendues de sables sorties de l'eau,
y compris les célèbres joyaux bien connus que sont
Koh Phangan, Koh Tao, Koh Samui et les douzaines
d'îlots du parc national marin d'Ang Thong. Si désirez
partir à la recherche de poissons avec votre appareil
photo, vous ne trouverez pas de meilleure région en
Thaïlande. Le parc national de Khao Sok est un rêve
en vert avec une végétation tropicale et des eaux peu
fréquentées.

Surat Thani & Ang Thong Marine National Park

Zur Provinz Surat Thani gehören Inseln wie Sand am
Meer, darunter die bekannten maritimen Juwelen Koh
Phangan, Koh Tao und Koh Samui und die dutzenden
Eilande des Ang Thong Meeresnationalparks. Wer mit
der Kamera auf Fisch-Safari gehen möchte, findet
kaum eine bessere Region in Thailand. Einen Traum
in Grün mit tropischer Vegetation und einsamen
Gewässern stellt der Khao Sok Nationalpark dar.

Khao Na Nai Luang, Surat Thani

Surat Thani & Ang Thong Marine National Park

La provincia de Surat Thani incluye islas como la de "arena junto al mar", incluyendo las famosas joyas marítimas de Koh Phangan, Koh Tao y Koh Samui y las docenas de islotes del parque nacional marino de Ang Thong. Si quieres ir a un safari de peces con tu cámara, difícilmente encontrarás una región mejor en Tailandia. El parque nacional Khao Sok es un sueño natural con vegetación tropical y aguas solitarias.

Surat Thani & Ang Thong Marine National Park

Le isole della provincia di Surat Thani sono numerosissime. Tra di esse ci sono anche le perle marine di Koh Phangan, Koh Tao e Koh Samui, nonché le decine di isolotti del Parco nazionale marino di Ang Thong. Non c'è posto migliore di questo in Thailandia per i fotografi appassionati di vita marina. Il Parco nazionale di Khao Sok offre uno spettacolo unico, immerso nel verde della vegetazione tropicale e con bacini d'acqua solitari.

Surat Thani & Ang Thong Marine National Park

Bij de provincie Surat Thani horen eilanden zoals zand bij de zee hoort, waaronder bekende maritieme juweeltjes als Koh Phangan, Koh Tao en Koh Samui en de tientallen eilandjes van het National Marine Park Ang Thong. Wie met zijn camera op een vissafari wil, vindt nergens in Thailand een betere regio. Het nationale park Khao Sok is een droom in groen met tropische begroeiing en stille wateren.

Khao Sok National Park, Surat Thani

Rafflesia kerrii, Khao Sok National Park, Surat Thani

Khao Sok National Park

Khao Sok National Park covers an area of 739 km^2 with dense rainforest, rushing waterfalls and idyllic Lake Cheow Lan, which measures over 160 km^2. An exciting boat trip of several hours on the beautiful water between the limestone rocks is worthwhile in order grasp the whole beauty of the park. Tapirs and elephants roam between green scrub and trees as tall as houses, and on the ground grows the world's largest flowering plant, the Bua Phut. It is the main attraction in Khao Sok, as its impressive flowers grow up to 1 m in size and can only be admired between October and December.

Parc national de Khao Sok

Le parc national de Khao Sok couvre une superficie de 739 km^2. On y trouve une forêt tropicale dense, des chutes d'eau rapides et un lac idyllique, le Cheow Lan, de plus de 160 km^2. Afin de pouvoir saisir toute la beauté du parc, rien de mieux que de réaliser une excursion de plusieurs heures en bateau entre les falaises calcaires. Des tapirs et des éléphants errent entre des arbres aussi hauts que des maisons et des broussailles vertes, et la plus grande plante à fleurs du monde, la Bua Phut, pousse sur le sol. C'est la véritable attraction de Khao Sok : ses fleurs impressionnantes, qui peuvent atteindre jusqu'à un mètre de hauteur, ne peuvent être admirées qu'entre octobre et décembre.

Khao Sok Nationalpark

Auf einer Fläche von 739 km^2 zeigen sich im Khao Sok Nationalpark dichter Regenwald, rauschende Wasserfälle und der idyllische, über 160 km^2 messende Cheow Lan-See. Auf dem wunderschön zwischen Kalksteinfelsen gelegenen Gewässer lohnt sich eine spannende mehrstündige Bootsfahrt, um die ganze Schönheit des Parks erfassen zu können. Zwischen haushohen Bäumen und grünem Gestrüpp streifen Tapire und Elefanten umher und am Boden wächst die größte Blütenpflanze der Welt, die Bua Phut. Sie ist die eigentliche Attraktion im Khao Sok, da ihre imposanten Blüten bis zu 1 m Größe heranwachsen und nur zwischen Oktober und Dezember zu bestaunen sind.

Khao Sok National Park, Surat Thani

Parque nacional Khao Sok

El parque nacional de Khao Sok tiene una superficie de 739 km², con un denso bosque pluvial, cascadas y el idílico lago de Cheow Lan, de más de 160 km². En el agua hermosa entre las rocas de piedra caliza, merece la pena hacer un viaje en barco emocionante de varias horas, con el fin de poder captar toda la belleza del parque. Los tapires y los elefantes deambulan entre árboles tan altos como casas y matorrales verdes, y la planta con flores más grande del mundo, la Bua Phut, crece en el suelo. Es el verdadero atractivo de Khao Sok, ya que sus impresionantes flores crecen hasta 1 m de tamaño y sólo se pueden admirar entre octubre y diciembre.

Parco nazionale di Khao Sok

Con la sua fitta foresta pluviale, le cascate scroscianti e l'idilliaco lago di Cheow Lan, di oltre 160 km², il Parco nazionale di Khao Sok si estende su una superficie di 739 km². Per poter cogliere tutta la bellezza del parco vale la pena fare un'emozionante gita in barca di alcune ore sulle bellissime acque tra le rocce calcaree. Tapiri ed elefanti vagano tra alberi alti come case e arbusti verdi; qui cresce la rafflesia, la più grande pianta a fiore del mondo. Questa è la vera attrazione di Khao Sok, poiché i suoi fiori impressionanti, che possono essere ammirati tra ottobre e dicembre, crescono fino a 1 m.

Nationaal park Khao Sok

Het beschermde natuurgebied Khao Sok beslaat een oppervlakte van 739 km² met dicht regenwoud, snelstromende watervallen en het idyllische Cheow Lan-meer van meer dan 160 km². Het is de moeite waard om een interessante boottocht van enkele uren te maken over het prachtige, tussen kalksteenrotsen gelegen water om zo de schoonheid van het gebied volledig in u op te kunnen nemen. Tapirs en olifanten zwerven rond tussen bomen zo hoog als huizen en groen struikgewas, en 's werelds grootste bloeiende plant, de bua phut, groeit op de grond. Dat is de echte attractie in Khao Sok, want de indrukwekkende bloemen kunnen wel 1 meter groot worden en zijn alleen tussen oktober en december te bewonderen.

Lamai Beach, Koh Samui

Fishing boats, Koh Samui

Koh Samui

Only a few inhabitants on Koh Samui still feed their families by fishing. However, their colourful boats are always a welcome eye-catcher. Thailand's third largest island lives mainly from tourism and has many dream beaches, however you will you rarely have these to yourself. Nevertheless, there is no shortage of idyllic locations and romantic spots. The beach of Chaweng on the west coast is beautiful, exceptionally popular and rich in bars and restaurants. Hardly less desirable with its snow-white sand and coconut palms swaying in the wind is the several kilometres long Lamai beach further south.

Koh Samui

Il ne reste plus que quelques pêcheurs à Koh Samui, mais leurs bateaux colorés continuent d'attirer l'attention. La troisième plus grande île de Thaïlande vit principalement du tourisme et possède de nombreuses plages de rêve, dont on profite rarement seul. Néanmoins, les lieux paradisiaques et romantiques ne manquent pas. La plage de Chaweng, sur la côte ouest, est très belle, exceptionnellement populaire et riche en bars et restaurants. Celle de Lamai, plus au sud, n'est pas loin derrière, avec son sable blanc comme neige et ses cocotiers qui se balancent dans le vent.

Koh Samui

Nur noch wenige Einwohner auf Koh Samui ernähren ihre Familien vom Fischfang. Ihre bunten Boote sind jedoch immer ein willkommener Blickfang. Die drittgrößte Insel Thailands lebt vorwiegend vom Tourismus und verfügt über viele Traumstrände, die man allerdings nur selten für sich alleine hat. An paradiesischen Orten und romantischen Plätzen besteht trotzdem kein Mangel. Wunderschön, ausnehmend beliebt und reich an Bars und Restaurants ist der Strand von Chaweng an der Westküste. Kaum weniger begehrenswert mit seinem schneeweißen Sand und den sich im Wind wiegenden Kokospalmen zeigt sich der mehrere Kilometer lange Lamai-Strand weiter südlich.

Koh Samui

Koh Samui

Sólo quedan unas pocas personas en Koh Samui que alimentan a sus familias con la pesca. Sin embargo, sus coloridos barcos siempre llaman la atención. La tercera isla más grande de Tailandia vive principalmente del turismo y tiene muchas playas de ensueño, que rara vez se disfrutan en solitario. Sin embargo, no faltan lugares paradisíacos y románticos. La playa de Chaweng, en la costa oeste, es hermosa, muy popular y llena de bares y restaurantes. No menos deseable, con su arena blanca como la nieve y sus cocoteros que se mecen con el viento, es la playa de Lamai, de varios kilómetros de largo, más al sur.

Koh Samui

Sono pochi a Koh Samui a vivere ancora esclusivamente di pesca. Tuttavia, le loro barche colorate sono sempre uno spettacolo molto apprezzato. La terza isola più grande della Thailandia vive oggi principalmente di turismo e ha molte spiagge da sogno, che però raramente si possono ammirare in solitudine. Tuttavia, non mancano i luoghi paradisiaci e romantici. Bella, eccezionalmente popolare e ricca di bar e ristoranti è la spiaggia di Chaweng, sulla costa occidentale. Non meno attraente è più a sud la spiaggia di Lamai, lunga diversi chilometri, con la sua sabbia bianca e le palme da cocco che ondeggiano al vento.

Koh Samui

Nog maar weinig bewoners van Koh Samui onderhouden hun gezinnen met de visvangst. Hun kleurige boten vormen echter altijd een welkome blikvanger. Het op twee na grootste eiland van Thailand leeft voornamelijk van het toerisme en beschikt over veel droomstranden, die men zelden voor zichzelf heeft. Toch is er geen tekort aan paradijselijke plekken en romantische plaatsen. Mooi, uitzonderlijk populair en rijk aan bars en restaurants is het strand van Chaweng aan de westkust. Weinig minder populair is het kilometerslange, verder naar het zuiden gelegen Lamai-strand met zijn sneeuwwitte zand en in de wind wuivende kokospalmen.

Koh Samui

Koh Samui

Wat Phra Yai on Koh Samui

Colourful shimmering sea, tropical coconut palms and world-famous beaches are the things that Koh Samui offers perfectly. Besides the prominent natural treasures, the island has important cultural attractions, such as Wat Phra Yai. The temple, erected in 1972 with its 12 m high shining golden Buddha is considered to be one of the landmarks of the island. The figure on the tiny island of Koh Phan, which is connected to Koh Samui by a narrow causeway, rises up into the Thai sky via a long staircase.

Wat Phra Yai sur Koh Samui

La mer colorée et chatoyante, les cocotiers tropicaux et les plages mondialement connues ne sont pas les seuls attraits de Koh Samui. Outre ses trésors naturels, l'île possède d'importantes attractions culturelles telles que le Wat Phra Yai. Le temple, avec son Bouddha doré de 12 m de haut érigé en 1972, est considéré comme l'une des réjouissances principales de l'île. Sur la petite presqu'île de Koh Phan, reliée à Koh Samui par une étroite bande de terre, on peut atteindre la statue qui s'élève dans le ciel thaïlandais par un long escalier.

Wat Phra Yai auf Koh Samui

Farbenprächtig schimmerndes Meer, tropische Kokospalmen und weltberühmte Strände sind nur die eine Klaviatur, die Koh Samui perfekt spielt. Neben den prominenten Naturschätzen verfügt das Eiland über wichtige kulturelle Attraktionen wie den Wat Phra Yai. Der Tempel mit seinem 12 m hohen, 1972 errichteten und goldfarben erstrahlenden Buddha gilt als eines der Wahrzeichen der Insel. In den thailändischen Himmel ragt die über eine lange Treppe erreichbare Figur auf der zwergenhaften Insel Koh Phan, die durch einen schmalen Damm mit Koh Samui verbunden ist.

Wat Phra Yai, Koh Samui

Wat Phra Yai en Koh Samui

El colorido y resplandeciente mar, los cocoteros tropicales y las playas de fama mundial son sólo algunos de los registros que la isla de Koh Samui domina a la perfección. Además de los destacados tesoros naturales, la isla cuenta con importantes atracciones culturales como el templo Wat Phra Yai. Este cuenta con 12 m de altura, posee un buda de color dorado erigido en 1972 considerado uno de los monumentos de la isla. La figura de la isla enana de Koh Phan, que está conectada con Koh Samui por una estrecha presa, se eleva hacia el cielo tailandés a través de una larga escalera.

Wat Phra Yai a Koh Samui

Il mare splendente e dalle tonalità accese, le palme da cocco e le spiagge famose in tutto il mondo sono solo alcune delle attrazioni di Koh Samui. Oltre ai famosi tesori naturali, l'isola ha importanti attrazioni culturali come il Wat Phra Yai. Il tempio, con il suo Buddha dorato alto 12 m eretto nel 1972, è considerato uno dei punti di riferimento dell'isola. Sulla minuscola isola di Koh Phan, collegata a Koh Samui da una stretta diga, si erge nel cielo thailandese una figura alta 12 m, a cui si accede attraverso una lunga scala.

Wat Phra Yai op Koh Samui

Een glinsterende, prachtig gekleurde zee, tropische kokospalmen en wereldberoemde stranden zijn niet het enige waar Koh Samui op kan bogen. Naast prominente natuurschatten beschikt het eiland over belangrijke culturele bezienswaardigheden zoals de Wat Phra Yai. Deze tempel met zijn 12 meter hoge, in 1972 opgerichte goudkleurige Boeddha wordt beschouwd als een symbool van het eiland. Een hoog de Thaise hemel instekende figuur op het dwergachtige eiland Koh Phan, dat door een smalle dam met Koh Samui is verbonden, is te bereiken via een lange trap.

Koh Samui

Than Sadet River, Koh Phangan

Koh Phangan

Koh Phangan

Koh Phangan

15 km north of Koh Samui, the 125 km² island of
Koh Phangan emerges from the sea. A beautiful
appearance that, like its sister island, boasts charming
beaches and attractive coral reefs. This jewel in the
Gulf of Thailand became world-famous due to the
wild monthly full moon parties that have been taking
place there for years.

Koh Phangan

A 15 km al norte de Koh Samui, la gran isla de 125 km²
Koh Phangan emerge del mar. Un bonito aspecto que,
al igual que su isla hermana, tiene playas elegantes
y atractivos arrecifes de coral. La joya se hizo
mundialmente famosa en el Golfo de Tailandia debido
a las fiestas mensuales de luna llena que han tenido
lugar durante años.

Koh Phangan

Koh Phangan, grande île de 125 km², se trouve à 15 km
au nord de Koh Samui. Aussi belle que la première,
elle offre elle aussi des plages gracieuses et des récifs
coralliens attrayants. Elle est par ailleurs devenue
mondialement célèbre dans le golfe de Thaïlande
pour les fêtes mensuelles de la pleine lune qui y ont
lieu depuis des années.

Koh Phangan

15 km a nord di Koh Samui emerge dal mare l'isola
di Koh Phangan, che ha una superficie di 125 km².
Un'apparizione incantevole che, come l'altra isola, ha
spiagge graziose e barriere coralline tutte da scoprire.
Questo gioiello nel golfo di Thailandia è diventato
famoso in tutto il mondo per via delle vertiginose
feste di luna piena che si svolgono da anni ogni mese.

Koh Phangan

15 km nördlich von Koh Samui taucht die 125 km²
große Insel Koh Phangan aus dem Meer auf. Eine
hübsche Erscheinung, die wie ihre Schwesterinsel mit
anmutigen Stränden und attraktiven Korallenriffen
aufwarten kann. Zu Weltruhm gelangte das
Schmuckstück im Golf von Thailand durch die seit
Jahren monatlich stattfindenden rauschenden
Vollmond-Partys.

Koh Phangan

15 km ten noorden van Koh Samui rijst het 125 km²
grote eiland Koh Phangan op uit zee. Een mooie
verschijning die, net als het zustereiland, beschikt over
verleidelijke stranden en aantrekkelijke koraalriffen.
Het juweeltje in de Golf van Thailand werd
wereldberoemd door de vollemaanfeesten die al jaren
maandelijks plaatsvinden.

Ao Chaloklum Bay, Koh Phangan

Koh Nangyuan

Freedom Beach, Koh Tao

Fish, Koh Tao

Diving area Koh Tao

Divers in particular make their way to the island of Koh Tao, whose name means turtle island. The marine habitat around the island was once rich in sea turtles. Today, this animal species, which moves amazingly gracefully through the water, is being reintroduced to the region. Abundant fish of all colours populate the coral reefs off the coast. The collar anemone fish hides from its predators between the tentacles of the anemone swaying in the current, while other fish prefer the protection of a shoal.

Zone de plongée de Koh Tao

Ce sont surtout les plongeurs qui visitent Koh Tao. Son nom signifie « île aux tortues », car elle était autrefois riche de cette espèce animale, qui se déplace avec une élégance étonnante dans l'eau ; aujourd'hui moins communes, les tortues marines ont tout de même été réintroduites dans la région. Par ailleurs, des poissons de toutes les couleurs peuplent les récifs coralliens au large de la côte. Le poisson-clown à collier se cache de ses prédateurs entre les tentacules des anémones se balançant dans le courant, tandis que les autres poissons préfèrent la protection d'un banc.

Tauchrevier Koh Tao

Insbesondere Taucher machen sich auf den Weg zur Insel Koh Tao, deren Name Schildkröteninsel bedeutet. Einst war das maritime Habitat rund um das Eiland reich an Meeresschildkröten. Heute versucht man diese sich im Wasser erstaunlich elegant fortbewegende Tierart wieder in der Region anzusiedeln. Im Überfluss hingegen bevölkern Fische aller Couleur die Korallenriffe vor der Küste. Der Halsband-Anemonenfisch versteckt sich vor seinen Fressfeinden zwischen den sich in der Strömung wiegenden Tentakeln der Anemone. Während andere schuppige Gesellen lieber den Schutz eines Schwarms bevorzugen.

Pink Anemonefish, Chumphon Pinnacle, Koh Tao

Zona de buceo Koh Tao

Los buzos son los visitantes más comunes de la isla de Koh Tao, cuyo nombre significa isla de las tortugas porque en el pasado, el hábitat marino alrededor de la isla era rico en tortugas marinas. Hoy en día, esta especie animal, que se mueve con increíble elegancia en el agua, está siendo reintroducida en la región. En cambio, peces de todos los colores pueblan en abundancia los arrecifes de coral de la costa. El pez anémona se esconde de sus depredadores entre los tentáculos de la anémona que se balancea en la corriente. Mientras que otras especiesprefieren la protección de un banco de peces.

Immersioni a Koh Tao

L'isola di Koh Tao, il cui nome significa isola delle tartarughe, è meta soprattutto di sub. Un tempo l'habitat marino intorno all'isola era popolato dalle tartarughe marine. Oggi questa specie, che si muove in modo sorprendentemente elegante in acqua, viene reintrodotta nella regione. Numerosi pesci variopinti si muovono invece lungo le barriere coralline al largo della costa. Il pesce pagliaccio crociato si nasconde ai suoi predatori tra i tentacoli oscillanti dell'anemone, mentre altri compagni squamosi preferiscono la protezione del branco.

Duikgebied Koh Tao

Vooral duikers gaan op weg naar het eiland Koh Tao, waarvan de naam schildpaddeneiland betekent. Ooit was de zee rond het eiland rijk aan zeeschildpadden. Tegenwoordig wordt deze diersoort, die zich in het water verbazingwekkend elegant voortbeweegt, opnieuw in de regio geïntroduceerd. Vissen in alle mogelijke kleuren bevolken daarentegen de koraalriffen voor de kust in overvloed. De *Amphiprion perideraion* verstopt zich voor zijn jagers tussen de in de stroming wiegende tentakels van de anemoon. Terwijl andere geschubde maten liever de bescherming van een school opzoeken.

Ang Thong National Marine Park

Thale Nai, Koh Mae Kohd, Ang Thong Marine National Park

Ang Thong Marine National Park

Ang Thong Marine National Park includes 42 islands, most of them uninhabited. Characteristic of this 102 km² region are the wooded karst cones, numerous caves and lonely beaches. The island of Koh Mae Koh, which can be reached by ferry from Koh Samui, boasts the Emerald Lake, the charm of which can best be seen from a viewing platform.

Parque nacional marino de Ang Thong

El parque nacional marino de Ang Thong incluye 42 islas, la mayoría de ellas deshabitadas. Característicos de esta región de 102 km² son los conos kársticos boscosos, las numerosas cuevas y las playas solitarias. La isla de Koh Mae Koh, a la que se puede llegar en ferri desde Koh Samui, cuenta con el Lago Esmeralda, cuya belleza se puede ver mejor desde un mirador.

Parc national marin d'Ang Thong

Le parc national marin d'Ang Thong comprend 42 îles, pour la plupart inhabitées. Dans cette région de 102 km², on trouve des cônes karstiques boisés, de nombreuses grottes et des plages solitaires. L'île de Koh Mae Koh, que l'on peut atteindre par ferry au départ de Koh Samui, est notamment célèbre pour le lac Émeraude, dont on peut admirer toute la grâce depuis une plate-forme panoramique.

Parco nazionale marino di Ang Thong

Il Parco nazionale marino di Ang Thong comprende 42 isole, la maggior parte delle quali disabite. Caratteristici di questa regione di 102 km² sono i monti carsici boscosi, le numerose grotte e le spiagge solitarie. L'isola di Koh Mae Koh, raggiungibile in traghetto da Koh Samui, si vanta della presenza del lago di Smeraldo, la cui bellezza può venire apprezzata al meglio dalla piattaforma panoramica.

Ang Thong Meeresnationalpark

Zum Ang Thong Meeresnationalpark zählen 42, meist unbewohnte Inseln. Charakteristisch für diese 102 km² große Region sind die waldbedeckten Karstkegel, zahlreiche Höhlen und einsame Strände. Die mit der Fähre von Koh Samui erreichbare Insel Koh Mae Koh trumpft mit dem Emerald-See auf, dessen gesamte Grazie man am besten von einer Aussichtsplattform ermessen kann.

Ang Thong Marine nationaal park

Het beschermde zeegebied Ang Thong bestaat uit 42 eilanden, waarvan de meeste onbewoond zijn. Kenmerkend voor deze 102 km² beslaande regio zijn de met bos bedekte karstkegels, talloze grotten en verlaten stranden. Het vanaf Koh Samui met de veerboot bereikbare eiland Koh Mae Koh beschikt over een smaragdgroen meer, waarvan de schoonheid het best te bewonderen is vanaf een uitkijkplatform.

Phang Nga & Similan Islands

Koh Kai, Phang Nga Bay

Khao Phing Kann (James Bond Island), Ao Phang Nga National Park

Phang Nga & Similan Islands
The western side of the Malay Peninsula and Phang Nga Bay serve as a magnificent stage for a play, with unique island diamonds and beach pearls as actors. One of these desirable locations is called Koh Kai, hardly bigger than a football field. The most famous attraction of the region is the "James Bond Island" Khao Phing Kan and its surroundings, formed from bizarre limestone rocks.

Phang Nga & Similan Islands
El lado occidental de la península de Malasia y la bahía de Phang Nga sirven como un magnífico escenario para una obra de teatro con diamantes isleños únicos y perlas de playa como protagonistas. Una de estas perlas se llama Koh Kai, apenas más grande que un campo de fútbol. Las atracciones más famosas de la región son la "Isla James Bond" Khao Phing Kan y los alrededores formados por extrañas rocas calizas.

Phang Nga & Similan Islands
Le côté ouest de la péninsule malaise et la baie de Phang Nga sont pleines d'îles magnifiques et de plages somptueuses. Koh Kai, à peine plus grande qu'un terrain de football, est l'une de ces pépites. Les attractions les plus célèbres de la région sont la « James Bond Island » Khao Phing Kan et les étonnantes roches calcaires de ses environs.

Phang Nga & Similan Islands
Il lato occidentale della penisola malese e la baia di Phang Nga fungono da magnifico palcoscenico per una serie di isole eccezionalmente belle e di spiagge mozzafiato. Tra queste Koh Kai, appena più grande di un campo di calcio. Le attrazioni maggiori della regione sono la Khao Phing Kan, l'isola famosa per un film di James Bond, e i suoi dintorni caratterizzati da bizzarre rocce calcaree.

Phang Nga & Similan Islands
Die westliche Seite der Malaiischen Halbinsel und die Bucht von Phang Nga dienen als grandiose Bühne für ein Theaterstück mit einzigartigen Inseldiamanten und Strandperlen als Darsteller. Einer dieser Sehnsuchtsorte heißt Koh Kai, kaum größer als ein Fußballfeld. Die bekanntesten Attraktionen der Region sind die „James-Bond-Insel" Khao Phing Kan und die aus bizarren Kalkfelsen geformte Umgebung.

Phang Nga & Similan Islands
De westkant van het Maleise schiereiland en de Baai van Phang Nga dienen als een grandioos podium voor een toneelstuk met unieke eilanden en stranden als hoofdrolspelers. Een van die droombestemmingen heet Koh Kai en is nauwelijks groter dan een voetbalveld. De bekendste bezienswaardigheden van de regio zijn het "Bondeiland" Khao Phing Kan en de door bizarre kalksteenrotsen gevormde omgeving.

Phang Nga

Phang Nga Bay, Ao Phang Nga National Park

Ao Phang Nga National Park

This is one of the most beautiful landscapes in southern Thailand. Large areas of the 400 km² protected area are traversed by picturesque rivers and covered by water. The James Bond movie *The Man with the Golden Gun* with Roger Moore was filmed in this region, and along with the hundreds of rock needles rising steeply from the sea; this made Phang Nga Bay world-famous. Only on a boat trip can you experience its charm and uniqueness in full splendour. A visit to a floating village such as Ban Ko Panyee rounds off the program nicely. There are several families in this settlement built on stilts, living mainly from fishing.

Parc national d'Ao Phang Nga

Le parc est un des plus beaux sites du sud de la Thaïlande. De larges zones de l'aire protégée de 400 km² sont traversées par des rivières pittoresques et couvertes d'eau. *L'Homme au pistolet d'argent,* film de la saga *James Bond* avec Roger Moore, a été tourné dans cette région – où des centaines d'aiguilles de roche s'élèvent abruptement de la mer – rendant ainsi la baie de Phang Nga mondialement célèbre. Une promenade en bateau vous permettra mieux que tout d'en ressentir le charme et la splendeur. La visite d'un village flottant comme Ban Ko Panyee complètera harmonieusement le programme. Plusieurs familles vivent dans la colonie construite sur pilotis, principalement de la pêche.

Nationalpark Ao Phang Nga

Es ist eine der schönsten Landschaften Südthailands. Weite Flächen des 400 km² großen Schutzgebietes sind von malerisch wirkenden Flussläufen durchzogen und von Wasser bedeckt. Der in dieser mit hunderten aus dem Meer steilaufragenden Felsnadeln gespickten Region gedrehte James Bond-Film *Der Mann mit dem goldenen Colt* mit Roger Moore machte die Bucht Phang Nga weltberühmt. Nur bei einem Bootsausflug lassen sich der Charme und die Einzigartigkeit in voller Pracht erleben. Der Besuch eines schwimmenden Dorfes wie Ban Ko Panyee rundet das Programm harmonisch ab. In der auf Stelzen erbauten Siedlung leben diverse Familien, hauptsächlich vom Fischfang.

Ban Ko Panyee Fishing Village, Phang Nga Bay

Parque nacional Ao Phang Nga

Es uno de los paisajes más bellos del sur de Tailandia. Las extensas áreas de los 400 km² del área protegida están atravesadas por pintorescos ríos y cubiertas de agua. La película de James Bond *The Man with the Golden Gun* con Roger Moore, rodada en esta región con cientos de agujas de roca que se elevan abruptamente desde el mar, hizo famosa mundialmente a la bahía de Phang Nga. Sólo en un viaje en barco se puede experimentar el encanto y la singularidad en todo su esplendor. Una visita a un pueblo flotante como Ban Ko Panyee completa el programa. En el asentamiento construido sobre pilotes viven varias familias, principalmente de la pesca.

Parco nazionale di Ao Phang Nga

È uno dei paesaggi più belli del sud della Thailandia. Vaste zone dei 400 km² dell'area protetta sono attraversate da pittoreschi fiumi e accolgono splendenti bacini d'acqua. La baia di Phang Nga è diventata famosa in tutto il mondo grazie al film di James Bond *L'uomo con la pistola d'oro,* con Roger Moore, girato in questa regione caratterizzata da centinaia di rocce appuntite a picco sul mare. Solo con una gita in barca si può apprezzare appieno il fascino e l'unicità di questi luoghi. Una visita a uno dei villaggi galleggianti come quello di Ban Ko Panyee renderà completo il programma. Nel villaggio, costruito su palafitte, vivono soprattutto famiglie di pescatori.

Nationaal park Ao Phang Nga

Het is een van de mooiste landschappen van Zuid-Thailand. Grote delen van het 400 km² beslaande natuurgebied worden doorkruist door schilderachtige rivieren en zijn bedekt met water. De Bondfilm *The Man with the Golden Gun* met Roger Moore, die in deze regio met zijn honderden steil uit de zee oprijzende naaldrotsen werd opgenomen, maakte de Baai van Phang Nga wereldberoemd. Alleen tijdens een boottocht kunt u de charme en uniciteit in volle pracht ervaren. Een bezoek aan een drijvend dorp als Ban Ko Panyee rondt het programma harmonieus af. In de op palen gebouwde nederzetting wonen verschillende families, voornamelijk van de visserij.

Khao Ta Pu, Phang Nga Bay

Koh Kho Khao

Ko Kho Khao and Koh Yao Noi

Ko Kho Khao is located about 100 km north of Phuket. The island enchants with its tranquillity, seclusion and endless beaches. This is a place to relax and soak up some sun, because there are no special sights. If you want to be active, you can take a diving tour to the nearby Similan Islands. The 12 km long island of Koh Yao Noi rises out of Phang Nga Bay. This small island near Phuket still exudes an original charm, and the inhabitants go fishing or cultivate their coconut plantations. You can reach villages and forests and the uncrowded beaches on foot.

Ko Kho Khao et Koh Yao Noi

Ko Kho Khao est situé à environ 100 km au nord de Phuket. L'île enchante par sa tranquillité, son isolement et ses plages infinies. C'est un endroit pour se détendre et prendre le soleil, notamment parce qu'il ne présente pas d'autres curiosités. Il est néanmoins possible d'y réaliser une excursion de plongée aux îles Similan voisines. Dans la baie de Phang Nga s'élève l'île de Koh Yao Noi, longue de 12 km, qui dégage encore un charme tout à fait original. Les habitants vont à la pêche ou cultivent leurs plantations de cocotiers. Vous pouvez atteindre les villages, les forêts et les plages non surpeuplées à pied.

Ko Kho Khao und Koh Yao Noi

Ko Kho Khao liegt rund 100 km nördlich von Phuket. Die Insel bezaubert durch Ruhe, Abgeschiedenheit und endlose Strände. Ein Ort zum Relaxen und Sonnenbaden, denn Sehenswürdigkeiten existieren nicht. Wer aktiv werden möchte, unternimmt eine Tauchtour zu den nahen Similan-Inseln. In der Phang Nga-Bucht erhebt sich die 12 km lange Koh Yao Noi-Insel. Das kleine Eiland nahe Phuket versprüht noch einen durchaus ursprünglichen Reiz. Die Einwohner gehen auf Fischfang oder pflegen ihre Kokosplantagen. Zu Fuß gelangt man in Dörfer und Wälder und zu nicht überfüllten Stränden.

Koh Yao Noi, Phang Nga Bay

Ko Kho Khao y Koh Yao Noi

Ko Kho Khao se encuentra a unos 100 km al norte de Phuket. La isla fascina por su tranquilidad, su aislamiento y sus playas interminables. Un lugar para relajarse y tomar el sol, porque no hay vistas. Si se desea algo de actividad, se puede hacer una excursión de buceo a las cercanas islas de Similan. En la bahía de Phang Nga se levanta la isla de Koh Yao Noi, de 12 km de largo. La pequeña isla cerca de Phuket todavía destila un encanto bastante original. Los habitantes van a pescar o cultivan sus plantaciones de coco. Se puede llegar a pie a pueblos y bosques, pero no a las playas superpobladas.

Ko Kho Khao e Koh Yao Noi

Ko Kho Khao si trova a circa 100 km a nord di Phuket. Tranquillità, isolamento e spiagge infinite rendono affascinante quest'isola. Questo è il luogo ideale per il relax e la vita da spiaggia, poiché non ci sono posti da visitare. Per un po' di attività fisica sono indicate le immersioni nelle vicine isole di Similan. Nella baia di Phang Nga si trova l'isola di Koh Yao Noi, lunga 12 km. La piccola isola, non lontana da Phuket, ha ancora tutto il fascino dell'incontaminato. Qui gli abitanti vivono di pesca o delle piantagioni di cocco. È possibile raggiungere a piedi villaggi, foreste e spiagge non sovraffollate.

Ko Kho Khao en Koh Yao Noi

Ko Kho Khao ligt ongeveer 100 km ten noorden van Phuket. Het eiland betovert met zijn rust, afzondering en eindeloze stranden. Een plek om te ontspannen en te zonnen, want er zijn geen bezienswaardigheden. Als u actief wilt worden, kunt u een duiktocht maken bij de nabijgelegen Similan-eilanden. In de Baai van Phang Nga verrijst het 12 km lange eiland Koh Yao Noi. Het kleine eilandje in de buurt van Phuket ademt nog een originele charme. De bewoners gaan vissen of werken op hun kokosplantages. U kunt dorpen en bossen en niet te volle stranden te voet bereiken.

Pasai Beach, Koh Yao Noi, Phang Nga Bay

Cauliflower jellyfish, Similan Islands National Park

Tiger tail seahorse, Similan Islands National Park

Underwater World of the Similan Islands

The Similan Islands in the Andaman Sea off the west coast of Thailand are a heavenly diving area. Creatures of unbelievable and graceful form swim in the clear water. While the diameter of the cauliflower jellyfish can reach almost 1 m, the tiger tail seahorse grows to a maximum size of 20 cm.

El mundo submarino de las islas Similan

Las islas Similan, frente a la costa occidental de Tailandia, en el mar de Andamán, son una zona de buceo paradisíaca. Las criaturas vivientes nadan en el agua clara de forma incomprensible y elegante. Mientras que el diámetro del paraguas de la medusa coliflor puede alcanzar casi 1 m, el caballito de mar cola de tigre crece hasta un tamaño máximo de 20 cm.

Monde sous-marin des îles Similan

Les îles Similan, situées au large de la côte ouest de la Thaïlande dans la mer d'Andaman, sont une zone de plongée paradisiaque. Des créatures de formes incompréhensibles mais gracieuses nagent dans l'eau claire. Alors que le diamètre du parapluie de la méduse chou-fleur peut atteindre près de 1 m, l'hippocampe à queue tigrée mesure au maximum 20 cm.

Il mondo sottomarino delle isole di Similan

Le isole di Similan al largo della costa occidentale della Thailandia, nel mare delle Andamane, sono un paradiso per chi ama le immersioni. Nelle acque limpide vivono esseri dalle forme insolite e gracili. Mentre il diametro dell'ombrello delle meduse cavolfiore può raggiungere quasi 1 m, il cavalluccio marino a coda di tigre cresce fino a raggiungere una dimensione massima di 20 cm.

Unterwasserwelt der Similan-Inseln

Ein himmlisches Tauchrevier sind die vor der Westküste Thailands, in der Andamanen-See liegenden Similan-Inseln. Im klaren Wasser schwimmen Lebewesen von unfassbarer und graziler Gestalt. Während der Schirmdurchmesser der Blumenkohl-Qualle fast 1 m erreichen kann, wird das Tigerschwanz-Seepferdchen maximal 20 cm groß.

De onderwaterwereld van de Similan-eilanden

De Similan-eilanden voor de westkust van Thailand, in de Andamanse Zee, zijn een hemels duikgebied. In het heldere water zwemmen levende wezens met onbegrijpelijke en gracieuze vormen. Terwijl de diameter van de bloemkoolkwal bijna 1 meter kan bedragen, wordt het zeepaardje *Hippocampus comes* maximaal 20 cm groot.

Koh Miang, Similan Islands National Park

Koh Miang, Similan Islands National Park

Similan Islands

Koh Miang and Koh Payu belong to an archipelago of 11 islands 60 km from the coast, which was declared a Marine National Park in 1982. The 70 km² area with its outstanding underwater world is described not without reason as one of the best diving areas in the world. The visibility between the rocks and coral reefs is excellent. Nothing clouds the dive down to clownfish, stingrays and sea turtles. Back on land, fine sandy beaches and well-formed rocky bays attract you.

Îles Similan

Koh Miang et Koh Payu appartiennent à un archipel de 11 îles à 60 km de la côte, déclaré parc national marin en 1982. La zone de 70 km² est l'un des meilleurs secteurs de plongée au monde. La vue entre les rochers et les récifs coralliens est excellente, on y observe des poissons-clowns, des raies et des tortues de mer et rien n'obscurcit le voyage. À terre, les plages de sable fin et les baies rocheuses bien dessinées complètent l'idyllique tableau.

Similan-Inseln

Koh Miang und Koh Payu gehören zum aus insgesamt 11 Inseln bestehenden und 60 km von der Küste entfernt liegenden Archipel, das 1982 zum Meeresnationalpark erklärt wurde. Das 70 km² umfassende Areal mit seiner exzellenten Unterwasserwelt wird nicht umsonst als eines der besten Tauchreviere der Welt bezeichnet. Die Sicht zwischen den Felsen und Korallenriffen ist exzellent, nichts trübt den Ausflug hinunter zu Clownfischen, Stachelrochen und Seeschildkröten. Wieder an Land locken feinsandige Strände und wohlgeformte Felsbuchten.

Koh Payu, Similan Islands National Park

Islas Similan

Koh Miang y Koh Payu pertenecen al archipiélago de 11 islas a 60 km de la costa, declarado parque nacional marino en 1982. La zona de 70 km² con su excelente mundo submarino no es una de las mejores zonas de buceo del mundo por casualidad. La vista entre las rocas y los arrecifes de coral es excelente, nada nubla el viaje hacia los peces payaso, rayas y tortugas marinas. De nuevo en tierra, llaman la atención las playas de arena fina y bahías rocosas bien formadas.

Isole di Similan

Koh Miang e Koh Payu appartengono a un arcipelago formato da 11 isole a 60 km dalla costa, che è stato dichiarato parco nazionale marino nel 1982. Con il suo incredibile mondo sottomarino, non a torto l'area di 70 km² è riconosciuta a livello mondiale come una delle migliori zone per immersioni. Tra rocce e barriere coralline si gode di una vista eccellente, sempre tersa durante l'immersione verso il basso alla scoperta di pesci pagliaccio, razze e tartarughe marine. Fuori dall'acqua, spiagge di sabbia fine e piacevoli baie rocciose sono una vera attrazione.

Similan-eilanden

Koh Miang en Koh Payu behoren tot een archipel van elf eilanden op 60 km van de kust, die in 1982 werd uitgeroepen tot beschermd zeegebied. Het 70 km² grote gebied met zijn uitstekende onderwaterwereld wordt niet voor niets beschouwd als een van de beste duikgebieden ter wereld. Het zicht tussen de rotsen en koraalriffen is voortreffelijk, niets vertroebelt de tocht naar beneden naar clownvissen, pijlstaartroggen en zeeschildpadden. Ook aan land trekken fijne zandstranden en goed gevormde rotsachtige baaien bezoekers aan.

Phuket

Pansea Beach, Phuket

Rubber plantation, Phuket

Phuket

The largest and probably most popular holiday
island of Thailand is Phuket. The province of the
same name also includes various smaller islands
such as Koh Naga Noi or Koh Hai. In the past, people
lived from profitable tin production. Today, rubber
plantations and mainly tourism bring cash into the
coffers. Visitors from all over the world fall in love
with the picturesque bays and beaches. Phuket city,
with its Chinese influenced architecture, its nine-day
"Vegetarian Festival" and the urban oases full of
tropical greenery is also thrilling.

Phuket

Phuket, célèbre station balnéaire, est la plus grande et
probablement la plus populaire des îles thaïlandaises.
Mais la province du même nom comprend également
diverses îles plus petites, comme Koh Naga Noi
ou Koh Hai. Dans le passé, les gens y vivaient
de la production d'étain. Aujourd'hui, ce sont les
plantations d'hévéa mais surtout le tourisme qui
nourrissent les habitants. Les visiteurs du monde
entier tombent amoureux des baies et des plages
pittoresques. Mais ce ne sont pas les seuls attraits
de la région : Phuket City, avec son architecture
d'influence chinoise, le « Vegetarian Festival » de
neuf jours ou les oasis urbaines pleines de verdure
tropicale valent également le détour.

Phuket

Die größte und wohl beliebteste Ferieninsel
Thailands ist Phuket. Zur gleichnamigen Provinz
gehören ferner diverse kleinere Inseln wie Koh
Naga Noi oder Koh Hai. Früher lebten die Menschen
von der gewinnbringenden Zinnförderung. Heute
bringen Kautschukplantagen und hauptsächlich
der Tourismus Geld in die Kassen. Vor allem in die
malerischen Buchten und Strände verlieben sich die
Besucher aus aller Welt. Doch auch Phuket-Stadt
mit ihrer chinesisch beeinflussten Architektur, dem
neuntägigen „Vegetarischen Fest" und den urbanen
Oasen voller tropischem Grün, begeistert auf
ganzer Linie.

Phuket Coast

Phuket

Phuket es laisla de vacaciones más grande y probablemente la más popular de Tailandia. La provincia del mismo nombre también incluye varias islas más pequeñas como Koh Naga Noi o Koh Hai. En el pasado, la gente vivía de la producción rentable de estaño. Hoy en día, las plantaciones de caucho y principalmente el turismo ingresan dinero en las arcas. Visitantes de todo el mundo se enamoran de las pintorescas bahías y playas. Pero la ciudad de Phuket, con su arquitectura de influencia china, el "Festival Vegetariano" de nueve días y los oasis urbanos llenos de vegetación tropical también son emocionantes.

Phuket

L'isola più grande della Thailandia, e probabilmente la più popolare tra i turisti, è Phuket. La provincia omonima comprende anche varie isole minori come Koh Naga Noi e Koh Hai. In passato, la gente viveva qui della produzione di stagno. Oggi le risorse più importanti sono le piantagioni di gomma e soprattutto il turismo. I visitatori provenienti da tutto il mondo ammirano qui le baie e le spiagge pittoresche. Ma Phuket City, con la sua architettura cinese, il "Festival vegetariano", che dura nove giorni, e le oasi urbane piene di verde tropicale è una meta attraente sotto molteplici aspetti.

Phuket

Het grootste en waarschijnlijk populairste vakantie-eiland van Thailand is Phuket. De gelijknamige provincie omvat ook verschillende kleinere eilanden zoals Koh Naga Noi en Koh Hai. Vroeger leefden mensen van de winstgevende tinproductie. Tegenwoordig brengen rubberplantages en vooral het toerisme geld in het laatje. Bezoekers van over de hele wereld worden verliefd op de schilderachtige baaien en stranden. Maar de stad Phuket met zijn door de Chinezen beïnvloede architectuur, het negendaagse "Vegetarische Feest" en de stadse oases vol tropisch groen bekoort ook over de hele linie.

Pansea Beach, Phuket

Kata Beach and Karon Beach, Phuket

Beaches of Phuket

Phuket's most popular beaches are on the west coast, including Kata Beach, Karon Beach and Kalim Beach. In addition to offering first-class water sports, these friendly places offer numerous restaurants, cafés and bars, which make taking a break under the Thai sun even more relaxing.

Plages de Phuket

Les plages les plus populaires de Phuket, parmi lesquelles Kata Beach, Karon Beach et Kalim Beach, se trouvent sur la côte ouest. En plus d'offrir des sports nautiques de première classe, ces lieux comportent de nombreux restaurants, cafés et bars qui participent également de cette ambiance de détente sous le soleil thaïlandais.

Strände von Phuket

Die begehrtesten Strände Phukets breiten sich an der Westküste aus, darunter der Kata Beach, Karon Beach und der Kalim Beach. Neben dem Angebot von erstklassigen Wassersportmöglichkeiten bieten diese freundlichen Orte zahlreiche Restaurants, Cafés und Bars, die das Entspannen unter der Sonne Thailands zusätzlich versüßen.

Kalim Beach, Phuket

Playas de Phuket

Las playas más populares de Phuket están en la costa oeste, incluyendo Kata Beach, Karon Beach y Kalim Beach. Además de ofrecer deportes acuáticos de primera clase, estos acogedores lugares ofrecen numerosos restaurantes, cafés y bares, que además endulzan la tranquilidad bajo el sol tailandés.

Le spiagge di Phuket

Le spiagge più popolari di Phuket, tra cui Kata Beach, Karon Beach e Kalim Beach, si trovano sulla costa occidentale. Oltre alla possibilità di cimentarsi in sport acquatici di prima classe, in questi luoghi ameni si trovano numerosi ristoranti, caffè e bar che rendono ancora più piacevole il relax sotto il sole thailandese.

Stranden van Phuket

De populairste stranden van Phuket liggen aan de westkust, zoals Kata Beach, Karon Beach en Kalim Beach. Behalve dat ze uitstekende watersportfaciliteiten bieden, beschikken deze vriendelijke plaatsen over tal van restaurants, cafés en bars, die de ontspanning onder de Thaise zon nog aangenamer maken.

Phuket

Peacock

Phuket's fauna

Phuket's dream beaches are not its only assets. In the mountainous interior and in Sirinath National Park to the north, nature is in the foreground. While colourful fish roam in the tropical water, animals such as proud peacocks or delicate butterflies live on the mainland. If you wish to see these fluttering creatures in all their fantastic forms, visit the beautifully laid out butterfly garden near Phuket City.

La faune de Phuket

Il n'y a pas que des plages de rêve à Phuket. Dans l'intérieur montagneux de l'île comme dans le parc national Sirinath, au nord, la nature est au premier plan. Tandis que les poissons colorés nagent dans les eaux tropicales, des animaux comme le fier paon ou diverses espèces de papillons vivent sur le continent. Si vous désirez admirer les formes incroyables de ces derniers, rendez-vous au magnifique jardin aux papillons, près de la ville de Phuket.

Phukets Fauna

Nicht nur Traumstrände sind das Kapital von Phuket. Im bergigen Landesinneren und im nördlich gelegenen Sirinath-Nationalpark steht die Natur im Vordergrund. Während sich im tropischen Nass farbenreiche Fische tummeln, bewohnen das Festland Tiere wie der stolze Pfau oder der feingestaltete Schmetterling. Wer diese flatterhaften Gesellen in all ihren fantastischen Ausprägungen beobachten möchte, besucht den schön angelegten Schmetterlingsgarten nahe Phuket-Stadt.

Tropical butterfly

La fauna de Phuket

No sólo las playas de ensueño constituyen la capital de Phuket. En el interior montañoso del país y en el parque nacional de Sirinath, situado al norte, la naturaleza está en primer plano. Mientras los peces de colores se divierten en las aguas tropicales, animales como el orgulloso pavo real o la mariposa de forma fina conviven en el continente. Si desea observar a estos viajeros revoloteando en todas sus fantásticas formas, visite el hermoso jardín de mariposas cerca de la ciudad de Phuket.

La fauna di Phuket

Non sono solo le spiagge da sogno ad essere l'attrazione principale di Phuket. Nell'entroterra montuoso e nel Parco nazionale di Sirinath, a nord, è la natura a stare in primo piano. Mentre pesci colorati si immergono nelle acque tropicali, animali come pavoni e farfalle graziose vivono sulla terraferma. Per osservare da vicino questi esseri svolazzanti in tutte le loro forme fantastiche basta recarsi al bellissimo giardino delle farfalle, poco distante da Phuket.

De fauna van Phuket

Niet alleen droomstranden vormen het kapitaal van Phuket. In het bergachtige binnenland en in het nationale park Sirinath in het noorden staat de natuur op de voorgrond. Terwijl bontgekleurde vissen spelen in het tropische water, leven dieren zoals de trotse pauw en de fijn gevormde vlinder op het vasteland. Als u deze fladderende maatjes in al hun fantastische vormen wilt observeren, bezoek dan eens de prachtig aangelegde vlindertuin in de buurt van de stad Phuket.

Coastline, Phuket

Seascape, Phuket

BUDDHAS

Whatever region you travel to, they are always present: Buddha figures. Sometimes a few centimetres tall, sometimes dozens of metres into the sky. Sometimes exposed standing on a mountain, sometimes sitting in front of a temple, these figures honour Siddhartha Gautama, who was probably born in India in the 6th century B.C. and is considered to be the founder of Buddhism.

BOUDDHAS

Quelle que soit la région dans laquelle vous voyagez, les représentations de Bouddha sont toujours présentes. Elles mesurent quelques centimètres de haut ou des dizaines de mètres, sont exposées debout sur une montagne ou assises, devant un temple... Elles rendent hommage à Siddhartha Gautama, probablement né en Inde au VIᵉ siècle avant J.-C. et considéré comme le fondateur du bouddhisme.

BUDDHAS

In welche Region man auch reist, sie sind immer präsent: Buddhafiguren. Mal wenige Zentimeter groß, mal dutzende Meter in den Himmel ragend. Mal exponiert auf einem Berg stehend, mal sitzend vor einem Tempel. Damit geehrt wird Siddhartha Gautama, der wahrscheinlich im 6. Jahrhundert v. Chr. in Indien geboren wurde, und als Begründer des Buddhismus gilt.

BUDAS

Independientemente de la región a la que se viaje, las figuras de Buda siempre están presentes. A veces solo miden unos pocos centímetros de altura, a veces decenas de metros hacia el cielo. Unas veces de pie en una montaña, otras sentado frente a un templo. Así se rinde homenaje a Siddhartha Gautama, quien probablemente nació en la India en el siglo VI a.C. y es considerado el fundador del budismo.

BUDDHAS

Qualunque sia la regione in cui si voglia andare, le statue del Buddha sono sempre presenti: a volte di pochi centimetri di altezza, altre allungate verso il cielo per decine di metri, a volte in piedi su una montagna, altre sedute davanti a un tempio. Con queste figure si onora Siddhartha Gautama, il Buddha storico, probabilmente nato in India nel VI secolo a.C., considerato il fondatore del buddismo.

BOEDDHA'S

Naar welke regio u ook reist, ze zijn altijd aanwezig: Boeddhabeelden. Soms zijn ze maar een paar centimeter hoog, soms steken ze tientallen meters de lucht in. Soms staan ze op een berg, soms zitten ze voor een tempel. Het is een eerbetoon aan Siddhartha Gautama, die waarschijnlijk in de 6e eeuw voor Christus in India werd geboren en beschouwd wordt als de grondlegger van het boeddhisme.

Krabi, Koh Lanta & Phi Phi Islands

Railay Beach, Krabi

Railay Beach, Krabi

Koh Tub and Koh Poda

Krabi, Koh Lanta & Phi Phi Islands

The world-famous Phi Phi Islands and Koh Lanta are part of the province of Krabi. Characteristic are the bizarre limestone rocks and beaches, for example the incomparable Railay Beach, which can only be reached by boat. The Phi Phis are a place of striking beauty. The interaction between beach, nature and the underwater world could hardly be more harmonious.

Krabi, Koh Lanta & Phi Phi Islands

Las islas mundialmente famosas de Phi Phi y Koh Lanta pertenecen a la provincia de Krabi. Se caracterizan por sus extrañas rocas de piedra caliza y sus playas, como la de Railay Beach, a la que sólo se puede llegar en barco. Las islas Phi Phis constituyen un lugar mimado por una belleza sorprendente. La interacción entre las playas, la naturaleza y el mundo submarino no puede ser más armoniosa.

Krabi, Koh Lanta & Phi Phi Islands

Les célèbres îles Phi Phi et Koh Lanta appartiennent à la province de Krabi. On y trouve des roches calcaires et d'étonnantes plages, comme celle de Railay Beach, qui ne peut être atteinte que par bateau. Les îles Phi Phi semblent avoir été touchées par la grâce. Le mariage des plages, de la nature et du monde sous-marin ne pourrait difficilement être plus harmonieux.

Krabi, Koh Lanta & Phi Phi Islands

Le isole Phi Phi e Koh Lanta, famose in tutto il mondo, fanno parte della provincia di Krabi. Caratteristica di questi posti sono le bizzarre rocce calcaree e le spiagge, come quella di Railay, raggiungibile solo in barca, che non conosce eguali. Un luogo di una bellezza sorprendente sono le isole Phi Phi: l'interazione tra spiagge, natura e mondo sottomarino non potrebbe essere più armoniosa.

Krabi, Koh Lanta & Phi Phi Islands

Die weltbekannten Phi Phi Inseln und Koh Lanta gehören zur Provinz Krabi. Charakteristisch sind die in den Himmel ragenden bizarren Kalksteinfelsen und Strände, wie der beispiellose Railay Beach, der nur per Boot erreichbar ist. Ein von auffallender Schönheit verwöhnter Ort sind die Phi Phis. Harmonischer kann das Zusammenspiel zwischen Stränden, Natur und Unterwasserwelt kaum gelingen.

Krabi, Koh Lanta & Phi Phi Islands

De wereldberoemde Phi Phi-eilanden en Koh Lanta horen bij de provincie Krabi. Kenmerkend zijn de bizarre kalksteenrotsen die de lucht insteken en stranden, zoals de ongeëvenaarde Railay Beach, die alleen per boot te bereiken zijn. De Phi Phi-eilanden zijn gezegend met een opvallende schoonheid. Het samenspel tussen strand, natuur en onderwaterwereld kan bijna niet harmonieuzer zijn dan hier.

Phra Nang Beach, Railay, Krabi

Wat Tham Suea (Tiger Cave Temple), Krabi

Wat Tham Suea

Temples such as Wat Tham Suea, which are often erected in exposed places, represent a welcome change to beach life. The building, known and appreciated as the Temple of the Tiger Cave, guards a Buddha footprint inside, and up hardly countable steps you reach a seated statue of the "Awakened One", whose unique location offers an outstanding view over the province of Krabi.

Wat Tham Suea

Les sanctuaires tels que Wat Tham Suea, souvent érigés dans des endroits exposés, sont une alternative réussie à la plage. Le bâtiment, connu sous l'appellation de « Temple de la grotte du tigre », abrite l'empreinte d'un Bouddha ; en grimpant un nombre difficilement mesurable de marches, on arrive à une statue de Bouddha assis, dont l'emplacement unique offre une vue exceptionnelle sur la province de Krabi.

Wat Tham Suea

Eine gelungene Abwechslung zum Strandleben stellen die oft an exponierter Stelle errichteten Heiligtümer wie der Wat Tham Suea dar. Das als Tigerhöhlentempel bekannte und geschätzte Bauwerk beschützt in seinem Inneren einen Fußabdruck Buddhas und über kaum zu zählende Stufen erreicht man eine sitzende Statue des „Erwachten", deren einmaliger Standort einen hervorragenden Ausblick über die Provinz Krabi gewährt.

Buddha Statues, Ao Nang, Krabi

Wat Tham Suea

Los santuarios como el Wat Tham Suea, que a menudo se erigen en lugares expuestos, representan una alternativa brillante a la playa. El edificio, conocido y apreciado como el Templo de la Cueva del Tigre, protege la huella de un Buda en su interior y por unos escalones que apenas se pueden contar se llega a una estatua sentada del "Despierto", cuya ubicación única ofrece una vista excepcional de la provincia de Krabi.

Wat Tham Suea

I santuari come il Wat Tham Suea, spesso costruiti in luoghi esposti, rappresentano di certo un'alternativa alla vita balneare. L'edificio, conosciuto e apprezzato come il Tempio della Grotta della Tigre, conserva al suo interno un'impronta del Buddha. Da una rampa di gradini, di cui è difficile tenere il conto, si raggiunge una statua seduta del "Risvegliato", la cui posizione unica offre una vista eccezionale sulla provincia di Krabi.

Wat Tham Suea

Heiligdommen zoals Wat Tham Suea, die vaak op gevaarlijke plaatsen worden gebouwd, vormen een geslaagde afwisseling van het strandleven. Het als 'Tempel van de tijgergrot' bekende gebouw beschermt binnen een voetafdruk van Boeddha. Over een nauwelijks te tellen aantal treden bereikt u een zittend standbeeld van de "Ontwaakte", waarvan de unieke positie een uitstekend uitzicht biedt over de provincie Krabi.

Sa Morakot, Krabi

Krabis natural phenomena

The crystal clear, blue shimmering water of Sa Morakot, also known as the "Emerald Pool", attracts numerous visitors: a gem of rare shape hiding among lush greenery near the city of Krabi. Similarly unique are the caves and lakes scattered throughout the region, with rock formations reminiscent of mystical creatures.

Nature de Krabi

L'eau claire et bleue de Sa Morakot, également connue sous le nom de « Piscine d'émeraude », attire de nombreux visiteurs. C'est un bijou d'une forme rare se cachant au milieu d'une végétation luxuriante, près de la ville de Krabi. De même, les grottes et les lacs disséminés dans toute la région sont uniques, et présentent des formations rocheuses qui rappellent des créatures mystiques.

Krabis Naturphänomene

Das kristallklare, blau schimmernde Wasser des Sa Morakot, auch als „Emerald-Pool" bekannt, lockt zahlreiche Besucher an. Ein sich nahe der Stadt Krabi zwischen üppigem Grün versteckendes Kleinod von seltener Gestalt. Ähnlich einmalig erscheinen die überall in der Region verteilten Höhlen und Seen, die an mystische Wesen erinnernde Felsformationen zum Besten geben.

Sa Phra Nangl, Krabi

Fenómenos naturales del Krabis

Las cristalinas y azules aguas de Sa Morakot, también conocidas como la "Piscina Esmeralda", atraen a numerosos visitantes. Una joya de extraña forma que se esconde entre un exuberante verdor cerca de la ciudad de Krabi. Igualmente singulares son las cuevas y lagos diseminados por toda la región, con formaciones rocosas que recuerdan a criaturas místicas.

I fenomeni naturali a Krabi

L'acqua cristallina e scintillante di riflessi blu di Sa Morakot, conosciuta anche come la "piscina di smeraldo", è un gioiello di forma rara che si nasconde tra il verde lussureggiante vicino alla città di Krabi e attira numerosi visitatori. Altrettanto spettacolari sono le grotte e i laghi sparsi in tutta la regione, con formazioni rocciose che ricordano creature mistiche.

Natuurfenomenen van Krabi

Het kristalheldere, blauw glinsterende water van de Sa Morakot, ook wel bekend als de "Emerald Pool", trekt veel bezoekers. Een uitzonderlijk gevormd kleinood, verscholen tussen weelderig groen, in de buurt van de stad Krabi. Uniek zijn ook de grotten en meren die over de hele regio verspreid zijn, met rotsformaties die doen denken aan mystieke wezens.

Maya Bay, Koh Phi Phi Leh

Koh Phi Phi

Phi Phi Islands

From a bird's eye view you can see the almost perfect beauty of the Phi Phi Islands resting in the Andaman Sea. These include Koh Phi Phi Don and Koh Phi Phi Leh. The latter has rock walls hundreds of metres high, between which small bays and tiny beaches are squeezed. Maya Bay, which served as a picturesque backdrop for the Hollywood film *The Beach* with Leonardo DiCaprio, is internationally famous.

Îles Phi Phi

Vues d'en haut, les îles Phi Phi, dans la mer d'Andaman, sont d'une beauté parfaite, notamment Koh Phi Phi Don et Koh Phi Phi Leh. Cette dernière possède des falaises de plusieurs centaines de mètres de haut, entre lesquelles se serrent de minuscules plages et de petites baies, comme celle de Maya, devenue célèbre dans le monde entier pour avoir servi de toile de fond pittoresque au film hollywoodien *La Plage,* avec Leonardo DiCaprio.

Phi Phi Inseln

Aus der Vogelperspektive offenbart sich die nahezu vollkommene Wohlgestalt, der in der Andamanensee ruhenden Phi Phi Inseln. Dazu gehören Koh Phi Phi Don und Koh Phi Phi Leh. Letztere weist hunderte Meter hohe Felswände auf, zwischen die sich kleine Buchten und winzige Strände zwängen. Internationale Berühmtheit erlangte die dortige Maya-Bucht, die als malerische Kulisse für den Hollywoodfilm *The Beach* mit Leonardo DiCaprio diente.

Koh Phi Phi Leh

Archipiélago de Phi Phi

A vista de pájaro se puede ver la belleza casi perfecta de las islas Phi Phi descansando en el mar de Andamán. A este archipiélago pertenecen las islas Koh Phi Phi Don y Koh Phi Phi Leh. Esta última tiene paredes de roca de cientos de metros de altura, entre las que se aprietan pequeñas bahías y playas. La Bahía Maya, que sirvió de telón de fondo pintoresco para la película de Hollywood *La Playa* con Leonardo DiCaprio, se hizo famosa internacionalmente.

Isole Phi Phi

È dall'alto che si può ammirare appieno la bellezza quasi perfetta delle isole Phi Phi, dolcemente adagiate nel mare delle Andamane. Tra di queste ci sono Koh Phi Phi Don e Koh Phi Phi Leh. Quest'ultima ha pareti rocciose alte centinaia di metri, tra cui si incuneano piccole baie e spiagge. La baia di Maya è diventata famosa a livello internazionale per aver fatto da sfondo pittoresco al film hollywoodiano *La spiaggia* con Leonardo DiCaprio.

Phi Phi Eilanden

Vanuit vogelperspectief kunt u de bijna perfecte schoonheid van de Phi Phi-eilanden zien, die in de Andamanse Zee liggen. Tot deze eilandengroep behoren Koh Phi Phi Don en Koh Phi Phi Phi Leh. Deze laatste heeft rotswanden van honderden meters hoog, waartussen kleine baaien en minieme strandjes geklemd liggen. Internationale bekendheid kreeg de daar gelegen Maya Bay, die als schilderachtig decor diende voor de film *The Beach* met Leonardo DiCaprio.

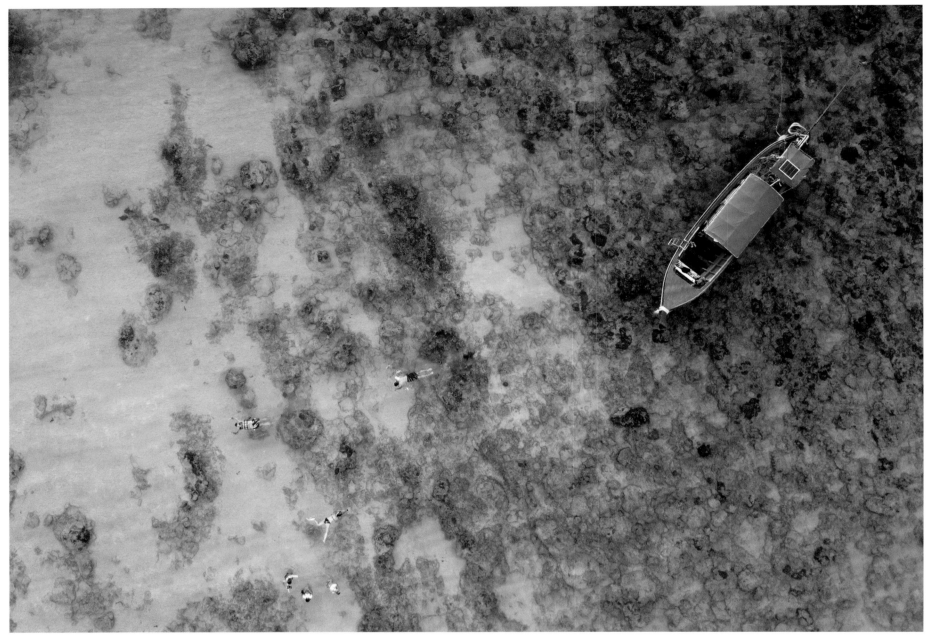

Koh Phi Phi

Taxi boats

The Phi Phi Islands are a feast for the eyes, as are the colourful, taxi boats floating in the clear water. Known locally as "long tail boats", they carry visitors swiftly and efficiently to any bay, no matter how remote, so it is not surprising that only a few desirable places in Thailand are still considered to be untouched and secluded.

Bateaux-taxis

Les îles Phi Phi sont un régal pour les yeux également grâce aux couleurs des bateaux-taxis, appelés localement « bateaux à longue traîne », qui conduisent rapidement les visiteurs dans toutes les baies, aussi éloignées soient-elles. Il n'est donc pas surprenant que seuls quelques lieux en Thaïlande soient encore considérés comme intacts et inhabités.

Taxiboote

Eine Augenweide sind nicht nur die Phi Phi Inseln, sondern auch die bunten, auf dem klaren Wasser liegenden Taxiboote, vor Ort „Long tail boats" genannt, die Besucher schnell und rasant zu jeder noch so abgelegenen Bucht chauffieren. So ist es nicht verwunderlich, dass nur noch wenige Sehnsuchtsorte in Thailand als unberührt und einsam gelten.

Long tail boats, Koh Phi Phi

Barcos taxi

No sólo las islas Phi Phi son un festín para la vista, sino también los coloridos barcos taxi aparcados en aguas claras, y localmente denominados "Long Tail Boats", que transportan a los visitantes a todas las bahías con un chófer, sin importar cuán remotas sean. Por lo tanto, no es de extrañar que sólo se consideren intactos y solitarios un par de paraísos en Tailandia.

Taxi-boat

Non sono solo le isole Phi Phi a essere una festa per gli occhi, ma anche i variopinti taxi-boat, le "barche dalla coda lunga" come che le chiamano i locali, in bella mostra sulle acque limpide, che conducono veloci i visitatori ovunque, anche nella baia più lontana. Non sorprende quindi che siano pochi i luoghi in Thailandia che possano essere ancora considerati incontaminati e solitari.

Taxiboten

Niet alleen de Phi Phi-eilanden zijn een lust voor het oog, maar ook de kakelbonte, op het heldere water dobberende taxiboten, die lokaal "long tail boats" worden genoemd en bezoekers snel naar alle verschillende baaien brengen, hoe afgelegen ook. Het is dus niet verwonderlijk dat nog maar weinig droombestemmingen op Thailand nog als ongerept en verlaten kunnen worden beschouwd.

Koh Lanta

The islands of Koh Lanta Yai and Koh Lanta Noi are situated south of the city of Krabi. The crescent-shaped and kilometre-long beach is only one of many palm-fringed dream beaches in the region. If you want to enjoy a tropical paradise to the fullest, romantic resorts and dreamy restaurants directly beside the sea invite you to stay.

Koh Lanta

Les deux îles Koh Lanta Yai et Noi s'étendent au sud de la ville de Krabi. La plage, en forme de croissant et longue d'un kilomètre, est représentative de la beauté des littoraux bordés de palmiers de la région. Les stations balnéaires romantiques et les restaurants de rêve au bord de la mer sont une invitation à profiter pleinement de ce paradis tropical.

Koh Lanta

Die beiden Inseln Koh Lanta Yai und Noi erstrecken sich südlich der Stadt Krabi. Der halbmondförmige und kilometerlange Long-Beach ist nur einer von vielen palmengesäumten Traumstränden der Region. Wer das tropische Paradies in vollen Zügen genießen möchte, den laden romantische Resorts und verträumte Restaurants direkt am Meer zum Bleiben ein.

Koh Lanta

Las dos islas, Koh Lanta Yai y Noi, se extienden al sur de la ciudad de Krabi. La playa, en forma de media luna y de un kilómetro de largo, es sólo una de las muchas playas de ensueño bordeadas de palmeras de la región. Quien desee disfrutar del paraíso tropical en los más completos complejos románticos y restaurantes de ensueño junto al mar, se sentirá invitado.

Koh Lanta

Le due isole Koh Lanta Yai e Noi si estendono a sud della città di Krabi. La spiaggia lunga chilometri a forma di mezzaluna, con il suo bordo di palme, è solo una dei tanti luoghi da sogno della regione. Chi voglia godersi al massimo il piacere di un paradiso tropicale come questo, numerosi resort romantici e ristoranti in riva al mare invitano a rimanere.

Koh Lanta

De twee eilanden Koh Lanta Yai en Koh Noi strekken zich uit ten zuiden van de stad Krabi. Het halvemaanvormige en kilometerslange Long Beach is slechts een van de vele met palmbomen omzoomde droomstranden in de regio. Als u met volle teugen wilt genieten van het tropische paradijs, nodigen romantische resorts en dromerige restaurants direct aan zee u uit voor een verblijf.

Long Beach, Koh Lanta

Diamond Cliff Beach, Koh Lanta

Nakhon Si Thammarat & Songkhla

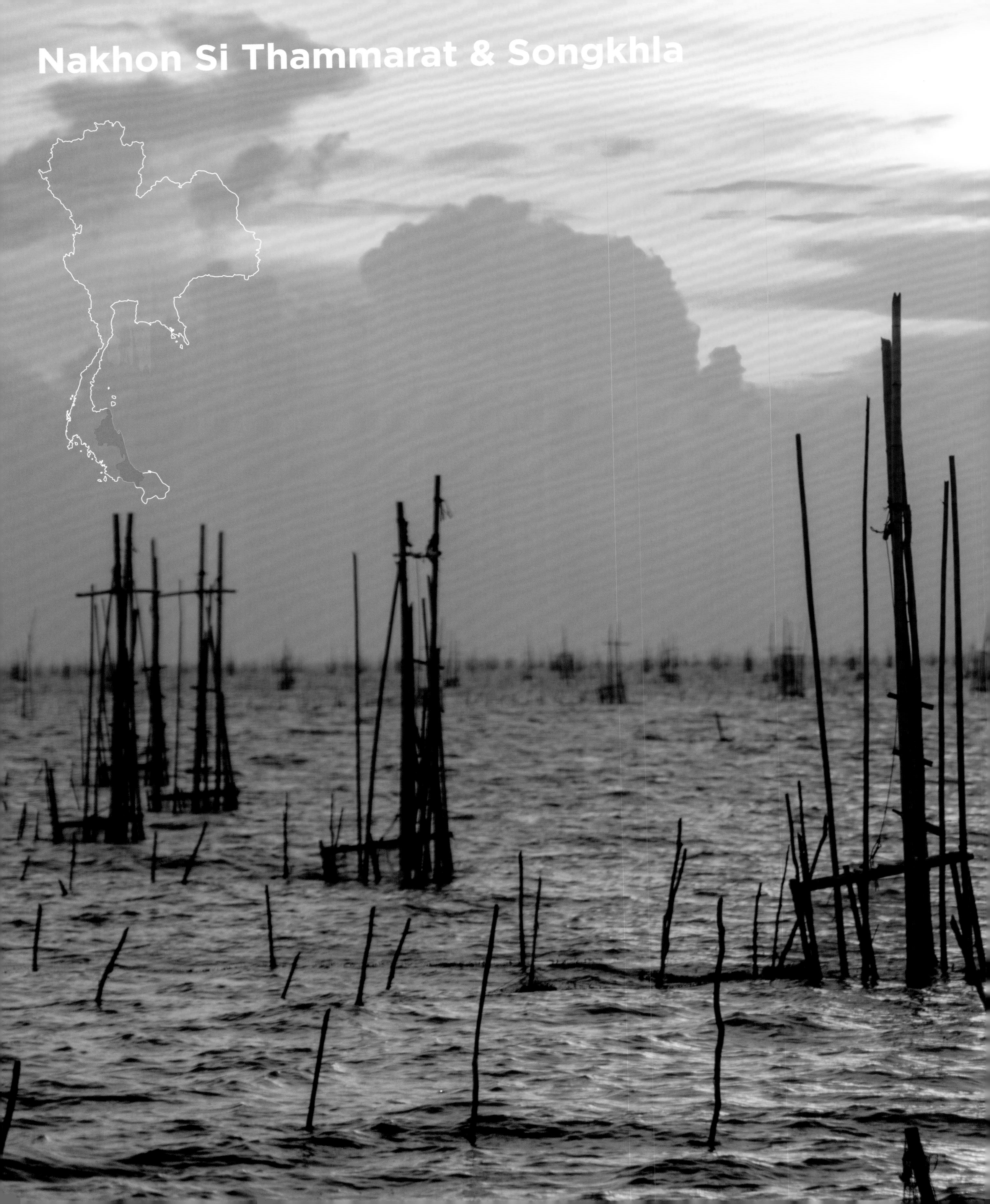

Bamboo fish trap, Songkhla Lake, Songkhla

Sunanta Waterfall, Khao Nan National Park, Nakhon Si Thammarat

Wat Phra Mahathat Woramaha Viharn, Nakhon Si Thammarat

Nakhon Si Thammarat & Songkhla

Nature and temple meet in the coastal province of Nakhon Si Thammarat. In the important Wat Phra Mahathat Woramaha Viharn, which is said to house a tooth relic of the Buddha, hundreds of white painted chedis are grouped around the more than 70 m high main chedi. Another important feature on the coast and on Lake Songkhla is fishing.

Nakhon Si Thammarat & Songkhla

La naturaleza y el templo se combinan en la provincia costera de Nakhon Si Thammarat. En el importante templo Wat Phra Mahathat Woramaha Viharn, que se supone que alberga una reliquia dentaria del Buda, cientos de chedis blancosse agrupan alrededor del Chedi principal de más de 70 m de altura. La pesca en la costa y en el lago Songkhla también desempeña un rol importante.

Nakhon Si Thammarat & Songkhla

Nature et spiritualité se rejoignent dans la province côtière de Nakhon Si Thammarat. Dans l'important Wat Phra Mahathat Woramaha Viharn, censé abriter une relique dentaire du Bouddha, des centaines de chedis blancs sont regroupés autour d'un chedi principal de plus de 70 m de haut. La pêche sur la côte et sur le lac Songkhla est également une activité courante dans la région.

Nakhon Si Thammarat & Songkhla

Natura e luoghi di spirito convivono nella provincia costiera di Nakhon Si Thammarat. Nell'importante tempio di Wat Phra Mahathat Woramaha Viharn, che ospita una reliquia dentale del Buddha, centinaia di chedi bianchi sono raggruppati intorno al chedi principale, alto oltre 70 m. La pesca lungo la costa e sul lago di Songkhla è un aspetto molto importante qui.

Nakhon Si Thammarat & Songkhla

Natur und Tempel treffen in der Küstenprovinz Nakhon Si Thammarat aufeinander. Im bedeutenden Wat Phra Mahathat Woramaha Viharn, der eine Zahnreliquie des Buddha beherbergen soll, gruppieren sich hunderte in weiß gehaltene Chedis um den über 70 m hohen Haupt-Chedi. Ebenfalls ein wichtiger Aspekt ist der Fischfang, der an der Küste und auf dem Songkhla-See betrieben wird.

Nakhon Si Thammarat & Songkhla

Natuur en tempel ontmoeten elkaar in de kustprovincie Nakhon Si Thammarat. In het belangrijke tempelcomplex Wat Phra Mahathat Woramaha Viharn, dat een tand van de Boeddha als relikwie zou huisvesten, zijn honderden wit gehouden chedi's gegroepeerd rond de meer dan 70 meter hoge hoofdchedi. Een ander belangrijk aspect is de visvangst, langs de kust en op het Songkhla-meer.

Buddha Statues Park, Nakhon Si Thammarat

Khao Kao Seng Beach, Songkhla

The legend of Nai Raeng

One of the most visited beaches in Songkhla province is Khao Kao Seng Beach, which has a story to tell. The so-called "Nai Raeng" rock, which is always decorated, was named after a governor who was once stranded there with his boat. On board, gold worth more than 900,000 Thai baht was stored, which Nai Raeng wanted to donate to honour Buddha in a ceremony. Due to his unfortunate situation, he missed the religious festival in Nakhon. It is said that deeply depressed, he then ordered his crew to put an end to his life and to hide the gold safely under the now revered rock.

La légende de Nai Raeng

L'une des plages les plus visitées de la province de Songkhla est celle de Khao Kao Seng Beach. Le rocher « Nai Raeng », toujours décoré, a été nommé d'après la légende du gouverneur du même nom, qui s'y serait échoué avec son bateau. On raconte qu'il transportait à bord pour 900 000 baht thaïlandais d'or, dont il comptait faire offrande à Bouddha lors d'une cérémonie religieuse à Nakhon. À cause de cet accident, il aurait manqué cette fête. Il aurait alors ordonné à son équipage, profondément déprimé, de mettre fin à sa vie et de cacher l'or en toute sécurité sous le rocher maintenant vénéré.

Die Legende von Nai Raeng

Einer der meistbesuchten Strände in der Provinz Songkhla ist der Khao Kao Seng Beach, der eine Geschichte zu erzählen hat. Der stets geschmückte, sogenannte „Nai Raeng"-Felsen wurde der Legende nach, nach dem gleichnamigen Gouverneur benannt, der dort einst mit seinem Boot strandete. An Bord lagerten 900 000 thailändische Baht in Gold, die Nai Raeng Buddha zu Ehren bei einer Zeremonie spenden wollte. Aufgrund seiner misslichen Lage verpasste er dieses religiöse Fest in Nakhon. Man sagt, dass er daraufhin tief deprimiert seiner Schiffsmannschaft befahl, seinem Leben ein Ende zu bereiteten und das Gold unter dem heute verehrten Felsen sicher zu verstecken.

Khao Kao Seng Beach, Songkhla

La leyenda de Nai Raeng

Una de las playas más visitadas de la provincia de Songkhla es la playa de Khao Kao Seng, que tiene una historia que contar. La roca llamada "Nai Raeng", permanentemente decorada, lleva el nombre de la leyenda del gobernador Nai Raeng , que una vez se quedó varado allí con su barco. A bordo se almacenaron 900.000 bahts tailandeses en oro, que Nai Raeng quería donar a en honor a Buda en una ceremonia. Debido a su situación, se perdió este festival religioso en Nakhon. Se cuenta que después ordenó a su tripulación, profundamente deprimida, que pusiera fin a su vida y escondiera el oro a salvo bajo la ahora venerada roca.

La leggenda di Nai Raeng

Una delle spiagge più visitate della provincia di Songkhla è quella di Khao Kao Seng, con una storia molto importante alle spalle. Lo scoglio chiamato "Nai Raeng", sempre decorato, prende il nome da un governatore che, secondo la leggenda, una volta vi si era arenato con la sua imbarcazione. A bordo vi si trovavano 900.000 baht thailandesi in oro che Nai Raeng voleva donare in onore del Buddha durante una cerimonia. A causa dell'incidente non riuscì ad andare alla festa a Nakhon. Si dice che allora ordinò all'equipaggio perplesso di porre fine alla sua vita e di nascondere l'oro sotto la roccia oggi venerata.

De legende van Nai Raeng

Een van de drukstbezochte stranden in de provincie Songkhla is Khao Kao Seng Beach, dat een verhaal te vertellen heeft. De altijd versierde Nai Raeng-rots is volgens de legende vernoemd naar de gelijknamige gouverneur, die daar ooit met zijn boot strandde. Aan boord lag voor 900.000 Thaise baht aan goud opgeslagen, dat Nai Raeng ter ere van een ceremonie aan Boeddha wilde schenken. Door zijn netelige situatie miste hij dit religieuze feest in Nakhon. Er wordt gezegd dat hij zijn bemanning, diep gedeprimeerd, opdracht gaf een einde aan zijn leven te maken en het goud veilig te verstoppen onder de nu vereerde rots.

Central Mosque, Hat Yai City, Songkhla

Trang & Satun

Koh Khai, Tarutao Marine National Park, Satun

Koh Lipe, Tarutao Marine National Park, Satun

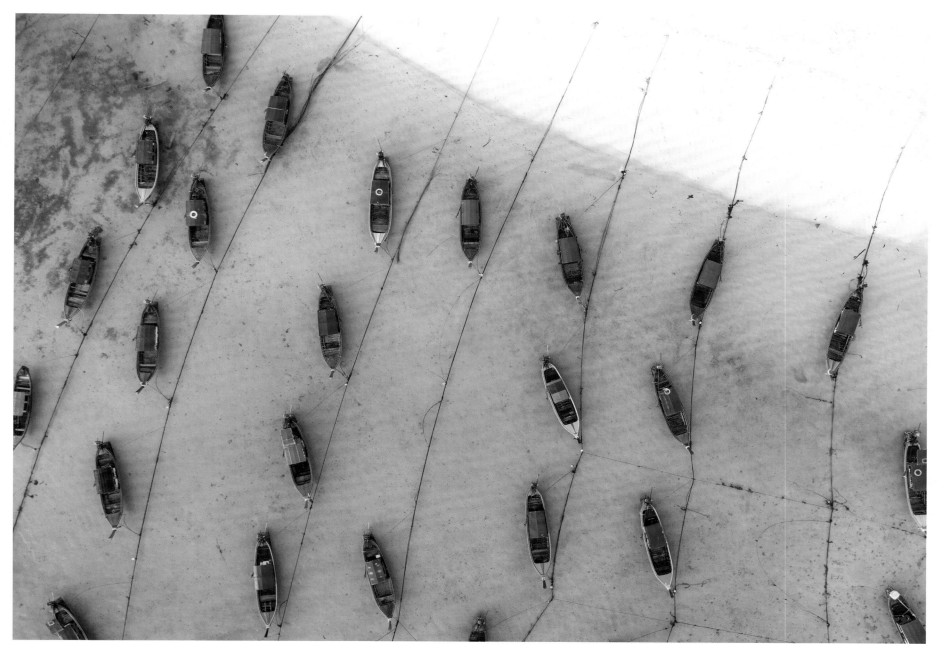

Koh Lipe, Tarutao Marine National Park, Satun

Trang & Satun

The provinces of Trang and Satun are located in the very south of Thailand on the border to Malaysia. A picture-book region with world-class diving areas and fabulous island landscapes. Rare creatures such as the hawksbill sea turtle are at home in the Tarutao Sea National Park in Satun Province.

Trang & Satun

Las provincias de Trang y Satun están situadas en el sur de Tailandia, en la frontera con Malasia. Una región de ensueño, con áreas de buceo de primera clase y fabulosos paisajes isleños. Las criaturas singulares, como la tortuga carey, viven en el parque nacional del Mar de Tarutao, en la provincia de Satun.

Trang & Satun

Les provinces de Trang et de Satun sont situées tout au sud de la Thaïlande, à la frontière avec la Malaisie. Une région de carte postale, avec des zones de plongée de classe mondiale et des paysages insulaires fabuleux. Des créatures rares, comme la tortue de mer à écailles, vivent dans le parc national marin de Tarutao, dans la province de Satun.

Trang & Satun

Le province di Trang e Satun si trovano nel sud della Thailandia, al confine con la Malaysia. La regione è come un libro illustrato, con aree per immersioni riconosciute a livello mondiale e favolosi paesaggi insulari. Nel Parco nazionale marino di Tarutao, nella provincia di Satun, vivono specie rare come la tartaruga embricata.

Trang & Satun

Ganz im Süden Thailands an der Grenze zu Malaysia erstrecken sich die Provinzen Trang und Satun. Eine Region wie aus dem Bilderbuch hinter der sich großartige Tauchreviere von Weltniveau und fabelhafte Insellandschaften verbergen. Seltene Geschöpfe, wie die Echte Karett-Meeresschildkröte, sind im Tarutao-Meeresnationalpark in der Provinz Satun zuhause.

Trang & Satun

De provincies Trang en Satun liggen in het uiterste zuiden van Thailand op de grens met Maleisië. Een regio als uit een prentenboek waarachter grandioze duikgebieden van wereldklasse en fantastische eilandlandschappen schuilgaan. Zeldzame wezens, zoals de karetschildpad, zijn thuis in het beschermde zeegebied Tarutao in de provincie Satun.

Koh Hin Ngam, Tarutao Marine National Park, Satun

Koh Hin Ngam, Tarutao Marine National Park, Satun

Koh Hin Ngam

Koh Hin Ngam is one of the smallest islands in the Tarutao Marine National Park. A special natural feature that can be found on the beach are the countless flattened pebbles. There are also some impressive stony outcrops. More than 50 islands belong to the protected area.

Koh Hin Ngam

Una de las islas más pequeñas del parque nacional marino de Tarutao se llama Koh Hin Ngam. Los innumerables guijarros aplastados que se pueden encontrar en la playa son una característica especial de la naturaleza. También hay algunos ejemplares impresionantes de piedra. Más de 50 islas se incluyen en el área protegida.

Koh Hin Ngam

L'une des plus petites îles du parc national marin de Tarutao s'appelle Koh Hin Ngam. Les innombrables galets aplatis que l'on peut trouver sur la plage sont une particularité de la nature. Plus de 50 îles appartiennent à cette zone protégée, dont certaines présentent également des étonnants pitons rocheux.

Koh Hin Ngam

Una delle isole più piccole del Parco nazionale marino di Tarutao è Koh Hin Ngam. Una particolarità che la natura qui offre sono gli innumerevoli ciottoli piatti che si possono trovare sulla spiaggia. Ma ci sono anche alcune rocce impressionanti. Più di 50 isole appartengono a quest'area protetta.

Koh Hin Ngam

Eine der kleinsten Inseln des Tarutao-Meeresnationalparks trägt den Namen Koh Hin Ngam. Eine von der Natur scheinbar gewollte Besonderheit sind die unzähligen dort am Strand zu findenden abgeflachten Kieselsteine. Einige imposante steinige Vertreter gibt es außerdem. Insgesamt über 50 Inseln gehören zum Areal des Schutzgebietes.

Koh Hin Ngam

Een van de kleinste eilanden in het beschermde zeegebied Tarutao heet Koh Hin Ngam. Een schijnbaar geliefde bijzonderheid van de natuur zijn de talloze afgeplatte kiezelstenen die daar op het strand te vinden zijn. Er zijn bovendien enkele indrukwekkende stenen vertegenwoordigers. Meer dan 50 eilanden behoren tot het areaal van het beschermde gebied.

Koh Kradan, Trang

Koh Kradan and Koh Mook

The islands in the province of Trang are a real feast for the eyes. Hills covered with tropical forest and fine sandy beaches, most of them empty, create a pleasant and relaxed atmosphere. Comfortable and tranquil resorts provide a touch of luxury. Deep caves and a spectacular underwater world guarantee an unforgettable stay.

Koh Kradan et Koh Mook

Les îles de la province de Trang sont un véritable régal pour les yeux. Les collines couvertes de forêt tropicale et les plages de sable fin, pour la plupart désertes, créent une atmosphère chaleureuse et détendue. Les stations balnéaires accueillantes et tranquilles offrent une touche de luxe. Des grottes profondes et un monde sous-marin spectaculaire finissent de garantir un séjour inoubliable.

Koh Kradan und Koh Mook

Eine wahre Augenweide sind die Inseln in der Provinz Trang. Mit tropischem Wald bedeckte Hügel und feinsandige meist leere Strände sorgen für eine wohlige und relaxte Atmosphäre. Gemütliche und beschauliche Resorts liefern einen Hauch von Luxus. Tiefe Höhlen und eine spektakuläre Unterwasserwelt garantieren einen unvergesslichen Aufenthalt.

Koh Mook, Trang

Koh Kradan y Koh Mook

Las islas de la provincia de Trang son un verdadero festival para la vista. Las colinas cubiertas de bosque tropical y playas de arena fina, la mayoría vacías, crean un ambiente acogedor y relajado. Los complejos hoteleros acogedores y tranquilos proporcionan un toque de lujo. Profundas cuevas y un espectacular mundo submarino garantizan una estancia inolvidable.

Koh Kradan e Koh Mook

Le isole della provincia di Trang sono un vero e proprio spettacolo della natura. Colline rivestite dalla foresta tropicale e spiagge di sabbia fine, la maggior parte delle quali deserte, creano un'atmosfera accogliente e rilassante. Resort tranquilli apportano un tocco di lusso allo scenario. Grotte profonde e uno spettacolare mondo sottomarino garantiscono un soggiorno indimenticabile.

Koh Kradan en Koh Mook

De eilanden in de provincie Trang zijn een lust voor het oog. Heuvels bedekt met tropische bossen en stranden met fijn zand, waarvan de meeste verlaten, zorgen voor een aangename en ontspannen sfeer. Gezellige en rustige resorts zorgen voor een vleugje luxe. Diepe grotten en een spectaculaire onderwaterwereld staan garant voor een onvergetelijk verblijf.

Golden Buddha at top of mountain, Trang

Islands of Andaman Sea

The islands of Trang

While there is a definite and quite pleasant level of activity on Koh Mook, the neighbouring island of Koh Kradan offers visitors peace and seclusion. If you want to enjoy this island beauty for a longer time, take a bungalow and enjoy your time to the full. However, should boredom arise the Andaman Sea has a rich variety of other islands to offer.

Îles de Trang

Bien que Koh Mook soit agréablement animée, l'île voisine de Koh Kradan offre aux visiteurs paix et isolement. Si vous désirez profiter plus longtemps de la beauté de cette île, l'idéal est de louer un bungalow. En cas d'ennui, la mer d'Andaman a un riche éventail d'autres îles à offrir.

Die Inseln von Trang

Während auf Koh Mook eine gewisse und durchaus angenehme Betriebsamkeit herrscht, schenkt die Nachbarinsel Koh Kradan den Besuchern vor allem Ruhe und Abgeschiedenheit. Wer länger in den Genuss dieser Inselschönheit kommen möchte, nimmt sich einen Bungalow und genießt die Auszeit in vollen Zügen. Sollte dennoch Langeweile aufkommen, hält die Andamanensee ein reichhaltiges Portfolio an weiteren Inseln bereit.

Koh Kradan, Trang

Las islas de Trang

Mientras que en Koh Mook hay una cierta actividad, bastante agradable , la isla vecina de Koh Kradan ofrece a los visitantes paz y aislamiento. Si se desea disfrutar de la belleza de esta isla por más tiempo, es recomendable alquilar un bungalow y disfrutar al máximo del tiempo libre. Si a pesar de todo el aburrimiento se presenta, el mar de Andamán tiene una rica cartera de otras islas que ofrecer.

Le isole di Trang

Mentre Koh Mook è caratterizzata da una certa attività, la vicina isola Koh Kradan offre ai visitatori pace e solitudine. Per chi voglia godersi al massimo la bellezza dell'isola per un periodo più lungo è consigliabile l'affitto di un bungalow. In caso di noia, tuttavia, il mare delle Andamane è ricco di isole da visitare.

De eilanden van Trang

Terwijl er op Koh Mook een zekere en volstrekt aangename bedrijvigheid heerst, biedt het naburige eiland Koh Kradan de bezoekers vooral rust en afzondering. Wie langer van de schoonheid van dit eiland wil genieten, huurt een bungalow en geniet met volle teugen van zijn time-out. Mocht de verveling dan toch toeslaan, dan heeft de Andamanse Zee een heel scala aan andere eilanden te bieden.

Chiang Mai	24
Chiang Rai	52
Mae Hong Son, Tak & Lamphun	68
Phayao, Nan & Phrae	84
Lampang	98
Sukhothai	112
Kamphaeng Phet, Phitsanulok & Phetchabun	126
Loei & Nong Khai	146
Udon Thani, Sakon Nakhon & Nakhon Phanom	162
Ubon Ratchathani	172
Nakhon Ratchasima, Buriram & Nakhon Nayok	184
Lopburi & Ayutthaya	196
Kanchanaburi	218
Bangkok	250
Samut Sakhon, Samut Prakan, Nakhon Pathom & Samut Songkhram	286
Chonburi	296
Rayong	320
Chantaburi & Trat	332
Ratchaburi & Phetchaburi	350
Prachuap Khiri Khan, Chumphon & Ranong	372
Surat Thani & Ang Thong Marine National Park	394
Phang Nga & Similan Islands	428
Phuket	450
Krabi, Koh Lanta & Phi Phi Islands	468
Nakhon Si Thammarat & Songkhla	492
Trang & Satun	504
	518

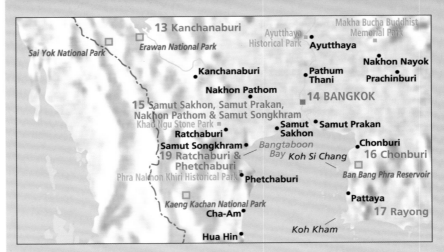

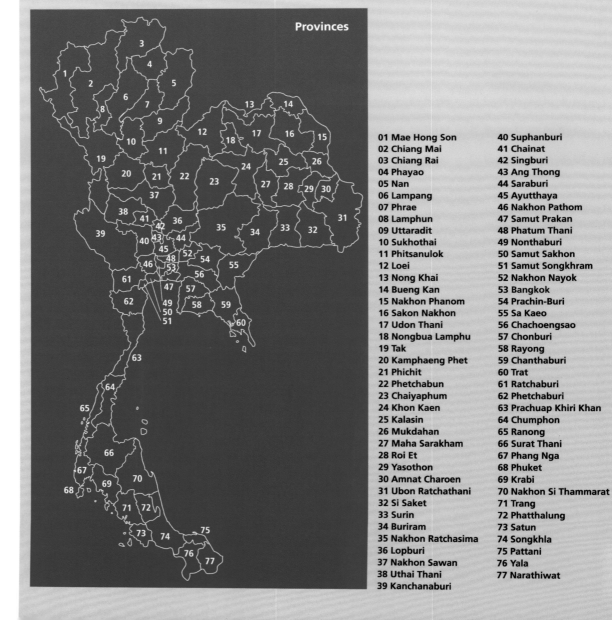

Provinces

01 Mae Hong Son	40 Suphanburi
02 Chiang Mai	41 Chainat
03 Chiang Rai	42 Singburi
04 Phayao	43 Ang Thong
05 Nan	44 Saraburi
06 Lampang	45 Ayutthaya
07 Phrae	46 Nakhon Pathom
08 Lamphun	47 Samut Prakan
09 Uttaradit	48 Phatum Thani
10 Sukhothai	49 Nonthaburi
11 Phitsanulok	50 Samut Sakhon
12 Loei	51 Samut Songkhram
13 Nong Khai	52 Nakhon Nayok
14 Bueng Kan	53 Bangkok
15 Nakhon Phanom	54 Prachin-Buri
16 Sakon Nakhon	55 Sa Kaeo
17 Udon Thani	56 Chachoengsao
18 Nongbua Lamphu	57 Chonburi
19 Tak	58 Rayong
20 Kamphaeng Phet	59 Chanthaburi
21 Phichit	60 Trat
22 Phetchabun	61 Ratchaburi
23 Chaiyaphum	62 Phetchaburi
24 Khon Kaen	63 Prachuap Khiri Khan
25 Kalasin	64 Chumphon
26 Mukdahan	65 Ranong
27 Maha Sarakham	66 Surat Thani
28 Roi Et	67 Phang Nga
29 Yasothon	68 Phuket
30 Amnat Charoen	69 Krabi
31 Ubon Ratchathani	70 Nakhon Si Thammarat
32 Si Saket	71 Trang
33 Surin	72 Phatthalung
34 Buriram	73 Satun
35 Nakhon Ratchasima	74 Songkhla
36 Lopburi	75 Pattani
37 Nakhon Sawan	76 Yala
38 Uthai Thani	77 Narathiwat
39 Kanchanaburi	

MYANMAR

LAOS

VIETNAM

2 Chiang Rai
Chiang Rai
Mae Kok
Doi Pha Tang
Mae Hong Son   Pai   *Doi Chiang Dao*   *Doi Phuwae*
Pai Canyon   Phayao   *Phu Langka Forest Park*
Phayao
1 Chiang Mai   Nan
Doi Inthanon National Park   Chiang Mai   Chae Son National Park   **4 Phayao, Nan & Phrae**
Doi Inthanon   Lamphun   Lampang   *Phae Mueang Phi Forest Park*
5 Lampang   Phrae
Mae Ping National Park
Uttaradit   Phu Phra Bat National Park   Sala Kaew Ku Sculpture Park   *Mekong*
3 Mae Hong Son, Tak & Lamphun   Nong Khai
Loei   **8 Loei & Nong Khai**   Nakhon Phanom
Sukhothai Historical Park   Udon Thani   *Nong Han*
6 Sukhothai   Phu Hin Rong Kla National Park   Phu Kradueng National Park   Nong Bua Lamphu   Sakhon Nakhon
Tak   Sukhothai   *Nan*   *Phu Pa Por Hill*
Nam Tok Pha Charoen National Park   Phitsanulok   Thung Salaeng Luang National Park   *Phu Thap Boek*   **9 Udon Thani, Sakon Nakhon & Nakhon Phanom**
Mae Sot   *Kamphaeng Phet Historical Park*   Phetchabun   Khon Kaen
Kamphaeng Phet   *Ping*
Khlong Lan National Park   Amnat Charoen   *Sam Phan Bok*
7 Kamphaeng Phet, Phitsanulok & Phetchabun   **THAILAND**   Chaiyaphum   *Pha Taem National Park*
Nakhon Sawan
Kwai   Uthai Thani   Nakhon Ratchasima   Buriram   Ubon Ratchathani
Huai Kha Khaeng Wildlife Sanctuary   Chai Nat   *Chao Phraya*   **10 Ubon Ratchathani**
Khao Chang Phueak   *Khuean Si Nakharin National Park*   Lopburi   **11 Nakhon Ratchasima, Buriram & Nakhon Nayok**   *Huai Luang Reservoir*
Thong Pha Phum National Park   **12 Lopburi & Ayutthaya**
13 Kanchanaburi   Ayutthaya   *Khao Yai National Park*
Kanchanaburi   Pathum Thani   Nakhon Nayok
15 Samut Sakhon, Samut Prakan, Nakhon Pathom & Samut Songkhram   Samut Sakhon   **14 BANGKOK**   Prachinburi
Samut Prakan
Samut Songkhram   Koh Si Chang   Chonburi   **16 Chonburi**
Phetchaburi   *Bangtaboon Bay*   *Ban Bang Phra Reservoir*
19 Ratchaburi & Phetchaburi   Pattaya   Chantaburi
17 Rayong   **18 Chanthaburi & Trat**
Hua Hin   Koh Kham   *Koh Samet*   Trat
Khao Sam Roi Yot National Park   Koh Mun
Kui Buri National Park   Koh Chang   CAMBODIA
Khao Fa Chi   Koh Mak
Prachuap Khiri Khan   *Khao Chong Krachok*   Koh Rang
Khao Lom Muak   Koh Kut
Andaman Sea

Gulf of Thailand

20 Prachuap Khiri Khan, Chumphon & Ranong
Koh Kai
Chumphon   *Koh Ngam Noi*
Khao Na Nai Luang   *Koh Nangyuan*
Koh Tao
Ranong   Koh Phangan
21 Surat Thani & Ang Thong Marine National Park   Koh Samui
Kho Kho Khao   *Khao Sok National Park*   Surat Thani
Similan Islands National Park   *Khao Nan National Park*
22 Phang Nga & Similan Islands   Nakhon Si Thammarat
Phang Nga   **24 Krabi, Koh Lanta & Phi Phi Islands**   Bang Pak Phraek
Ao Phang Nga National Park   Krabi
Phuket   **25 Nakhon Si Thammarat & Songkhla**
23 Phuket   Phatthalung
Koh Lanta   Trang   Songkhla
Koh Mook   **26 Trang & Satun**   *Songkhla*
Koh Kradan   *Tarutao Marine National Park*   Hat Yai
Koh Hin Ngam   Satun

Koh Lipe

VIETNAM

MALAYSIA

Index

Ampawa 292
Ancient City Muang Boran 288 f.
Andaman Sea, Islands 516
Ang Thong 199, 216
Ang Thong Marine National Park 396, 424 ff.
Ao Chaloklum Bay 416 f.
Ao Katueng Beach 341
Ao Khao Kwai 390 f.
Ao Nang 475
Ao Phang Nga National Park 431, 434
Ayutthaya 199, 204
Ayutthaya Historical Park 198 f., 206, 209 ff.,
    214 f.

Ban Bang Phra Reservoir 312 f., 314 f.
Ban Ko Panyee Fishing Village 435
Bang Pa 204 f.
Bang Saray 310 f. (beach), 311
Bangkok 4 f., 66, 67, 253, 276
Bangkok Chinatown 280 f.
Bangkok Grand Palace 258 f., 260 f.
Bangkok Skyline 256 f., 262 f.
Bangkok's Streets 272 f.
Bangtaboon Bay 358
Bhumibol Bridge 286 f.
Blue Pitta 352
Bong Piang Forest 36
Bryce whales 364 f.
Buddha Statues Park 496 f.
Buffalo Horn Beach 390 f.
Buriram 18, 190, 192 f.

Cauliflower jellyfish 444
Chae Son National Park 108
Chantaboon Old Town 338 f.
Chantaburi 334, 338, 340, 342
Chao Phraya River 253, 274 f., 282 f.
Cherry Blossom Flower 152 f.
Chiang Khan 147
Chiang Khong 58
Chiang Mai 26, 27, 30, 40 f., 66
Chiang Mai City 38
Chiang Rai 55
Chiang Rai City 56 f.
Chinatown Bangkok 280 f.
Chonburi 66, 298
Chumphon 375, 387
Chumphon Pinnacle 423
Culinary 284 f.

Damnoen Saduak 354 f.
Diamond Cliff Beach 490 f.
Doi Chiang Dao Mountain 42 f., 50 f.
Doi Inthanon Mountain 25
Doi Inthanon National Park 34 f.
Doi Pha Tang Mountain 52 f.
Doi Phu Kha National Park 84 f.
Doi Phuwae Viewpoint 156 f.

Elephants 26, 160 f., 246 f., 385
Erawan National Park 242 ff.
Erawan Waterfall 242, 244 f.

Freedom Beach 420 f.

Golden Triangle 64 f.
Grand Palace, Bangkok 258 f., 260 f.
Gulf of Thailand 340

Haew Narok Waterfall 22, 188 f.
Hat Bang Ben Beach 388 f.
Hat Yai City 502
Ho Kham Luang 28
Hua Hin 384 f.
Hua Hin City 372 f.
Huai Kha Khaeng Wildlife Sanctuary 82 f.
Huai Luang Reservoir 162 f.
Huay Mae Khamin Waterfall 230 f.

James Bond Island 431

Kaeng Krachan Dam 360 f., 366 f.
Kaeng Krachan National Park 352, 360, 362 f.,
    366 f.
Kalim Beach 457
Kamphaeng Phet 128 f.
Kamphaeng Phet Historical Park 129 ff.
Kanchanaburi 220
Karon Beach 456
Kata Beach 456
Khao Chang Phueak Mountain 228 f.
Khao Chong Krachok Mountain 380 f., 382 f.
Khao Fa Chi Viewpoint 392 f.
Khao Kao Seng Beach 498 f.
Khao Kho 142, 144 f.
Khao Lom Muak Mountain 376 f.
Khao Na Nai Luang 397
Khao Nan National Park 494
Khao Ngu Stone Park 356 f.

Khao Phing Kann 431
Khao Sam Roi Yot National Park 375, 378 f.
Khao Sok National Park 398 f., 400 f.
Khao Ta Pu 438 f.
Khao Yai National Park 22, 186 f., 189
Khlong Lan National Park 128
Khmer Temple 192
Khuean Si Nakharin National Park 230
Ko Nangyuan 418 f.
Koakho and Phu Tub Mountain 140 f.
Koh Chang 332 f.
Koh Hin Ngam 508 f.
Koh Kai 428 f.
Koh Khai 504 f.
Koh Kham 316 f., 318 f.
Koh Kho Khao 440
Koh Kood 8 f.
Koh Kradan 512, 517
Koh Kut 344 ff.
Koh Lanta 471, 486 f., 489, 491
Koh Lipe 506 f.
Koh Mae Kohd 426
Koh Mak 340 f., 348 f.
Koh Miang 446 f., 448
Koh Mook 512 f.
Koh Mun 323
Koh Ngam Noi 386 f.
Koh Payu 449
Koh Phangan 412, 414 ff.
Koh Phayam 390
Koh Phi Phi 480, 484 f.
Koh Phi Phi Don 482 f.
Koh Phi Phi Leh 12 f., 479, 481
Koh Rang 349
Koh Samet 328 ff.
Koh Samui 402, 404 ff., 408 ff.
Koh Si Chang 296 f., 300 f.
Koh Tao 394 f., 421 ff.
Koh Wai 335
Koh Yao Noi 440 ff.
Krabi 15, 21, 469 ff., 473 ff.
Kui Buri National Park 385
Kung Krabaeng 342 f.

Lam Nam Kok National Park 60 f.
Lamai Beach 402 f.
Lampang 101
Lamphun 68, 70, 74
Loei 66, 147 ff., 152, 157

Loi Krathong 66 f.
Lom Phu Kiew 98 f.
Long Beach, Koh Lanta 488 f.
Lopburi 197, 199 f., 202 f.
Lumpini Park 254 f.

Macaque 243
Mae Hong Son 69 f., 72 f.
Mae Kok River 55
Mae Ping National Park 74 f.
Mae Sot 76 f.
Mae Tha 102
Mae Ya Falls 35
Maeklong Railway Market 293
Makha Bucha Buddhist Memorial Park 184 f.
Man Dang Waterfall 137
Mangrove Field, The Golden 326 f.
Mangrove Forests 324 f.
Maya Beach 478 f.
Mekong River 58, 64 f., 146 ff., 154f.
Mon Bridge 222 f., 232 f.
Muang Tham 18
Myanmar 238

Nai Raeng rock 498
Nakhon Nayok 184
Nakhon Pathom 289
Nakhon Phanom 164, 168, 171
Nakhon Ratchasima 186 f., 189
Nakhon Si Thammarat 494 f., 497
Nam Tok Pha Charoen National Park 78 f.
Nan 85 f., 92
Noen Nang Phaya Viewpoint 340
Nong Han Lake 166 f.
Nong Khai 148, 154, 158 f.
Nong Nooch Tropical Botanical Garden 306 f.

Oriental Pied Hornbill 362

Pai Canyon 68 f.
Pang Na Bay 2, 10 f.
Pansea Beach 450 f., 454 f.
Pasai Beach 442 f.
Pattaya 299, 303, 306, 308
Pattaya City 298, 304 f.
Peacock 460
Pee Ta Khon 66
Pha Taem National Park 179 ff.
Phae Mueang Phi Forest Park 94 f.

Phang Nga 432 f.
Phang Nga Bay 428, 430 f., 434 ff., 438, 441 f.
Phanom Rung 193
Phayao 86 ff., 90 f.
Phayao Lake 90 f.
Phetchabun 127, 129, 140, 142, 144 f.
Phetchaburi 351 ff., 358 ff., 362 f., 366 ff.
Phi Phi Islands 471, 480
Phitsanulok 129, 132, 135 ff.
Phra Maha Chedi Chai Mongkol 370
Phra Nakhon Khiri Historical Park 353
Phra Nang Beach 472 f.
Phrae 86, 95 ff.
Phraya Nakhon Cave 375
Phraya River 262 f.
Phu Chi Fa Mountain 54
Phu Hin Rong Kla National Park 132 f., 136 ff.
Phu Kradueng Natonal Park 150 f.
Phu Langka Forest Park 87 ff.
Phu Lom Lo 152
Phu Pa Por Hill 149
Phu Phra Bat National Park 164 f.
Phu Thap Boek Mountain 126 f.
Phuket 450, 452, 454, 456 ff., 464 f.
Phuket Coast 453, 462 f.
Pink Anemonefish 423
Pink-Lipped Habenaria 136
Porcupine, Malayan or Himalayan 363
Prachuap Khiri Khan 373 ff., 377 ff., 383 ff.
Prang Sam Yot 200 f., 202
Prasat Mueang Tham 190 ff.
Prasat Sut Ja-Tum 308 f.
Pua District 92 f.

Rafflesia kerrii 400
Railay 473
Railay Beach 14 f., 468 ff.
Rama VIII Bridge 253
Ranong 375, 388, 390, 393
Ratchaburi 352, 354, 356
Ratchada 266 f.
Ratchadamri Road, Bangkok 264
Rayong 320 ff.
River Kwai 220
River Kwai Bridge 248 f.
Royal barge procession 66
Royal Pavilion Ho Kham Luang 28 f.

Sa Morakot 476
Sa Phra Nangl 477
Sai Yok National Park 234
Sakon Nakhon 164, 166, 170
Sala Kaew Ku Sculpture Park 158 f.
Sam Phan Bok 16 f., 174 ff., 178
Samut Prakan 286, 288 ff.
Samut Sakhon 289, 294 f.
Samut Songkhram 289, 292 f.
Sanctuary of Truth 308 f.
Sangkhlaburi 218 f., 223 f., 227, 233, 236, 238
Sattahip 316, 318 f.
Satun 505, 507 ff.
Si Racha 312, 315
Si Satchanalai Historical Park 115
Sikao 510 f.
Similan Islands 431, 444, 448
Similan Islands National Park 444 ff., 448 f.
Skytrain 252, 256 f.
Songkalia River 236 f.
Songkhla 493, 495, 498 f., 502
Songkhla Lake 493
Songkran 67
Spirit houses 102 f., 110 f.
Sukhothai Historical Park 6 f., 112, 114, 116, 118 ff., 125
Sukothai 115
Sunanta Waterfall 494
Surat Thani 396 ff., 400 f.1

Tak 68, 70 f., 76 f., 79 f.
Tarutao Marine National Park 505 ff.
Thai Elephant 246 f.
Thai-Burma Railway 235
Thale Nai 426
Thale Noi 501
Tham Khao Luang Cave 368 f.
Than Sadet River 412 f.
Third Thai Lao Friendship Bridge 168 f.
Thong Pha Phum National Park 228
Thung Salaeng Luang National Park 144 f.
Tiger Cave Temple 474
Tiger tail seahorse 445
Train Night Market Ratchada 266 f.
Trang 507, 511 ff., 516 f.
Trat 332, 334 ff., 341, 344 ff., 348 f.
Tropical butterfly 461
Tuk-tuk 277
Tulay Hill 80 f.

Tung Prong Thong 326 f.
Ubon Ratchathani 16, 172 f., 174
Udon Thani 163 ff.

Wang Kaew Waterfall 104 f.
Wat Arun 250 f.
Wat Ban Rai 370
Wat Benchamabophit 371
Wat Chai Mongkol 100
Wat Chai Wattanaram 210
Wat Chalermprakiat Prajomklao Rachanusorn
   101
Wat Chang Lom 115
Wat Chedi Luang 39
Wat Chiang Man 370
Wat Djittabhawan 302 f.
Wat Huay Pla Kang 59
Wat Jed Yod 66
Wat Khao Phra Khru 66
Wat Mahathat 112 ff., 116 f., 121 ff., 212 f.
Wat Manee Pai Son 77
Wat Muang 216 f.

Wat Phaphutthabat Phuphadaeng 106 f.
Wat Pho 268 ff.
Wat Phra Dhammakaya 264 f., 371
Wat Phra Kaew 129 f., 258 f.
Wat Phra Mahathat Woramaha Viharn 495
Wat Phra Si Sanphet 208 f.
Wat Phra Singh 38
Wat Phra Sorn Kaew 142 f.
Wat Phra Sri Rattana Mahathat 134 f., 196 f., 203
Wat Phra That Choeng Chum 170
Wat Phra That Doi Din Chi 76
Wat Phra That Doi Kham 370
Wat Phra That Doi Kong Mu 370
Wat Phra That Doi Suthep 48 f.
Wat Phra That Panom 171
Wat Phra That Suthon Mongkol Khiri 96 f.
Wat Phra Yai 371, 408 f.
Wat Phra Yai Ko Pan 370
Wat Phumin 86
Wat Plai Laem 370
Wat Ratchanatdaram 278 f.
Wat Rong Khun 62 f.

Wat Saket 371
Wat Si Chum 120
Wat Srisuphan 370
Wat Tha Khanung 239
Wat Tha Ton 44 f.
Wat Tham Heo Sin Chai 182 f.
Wat Tham, Sua 240 f.
Wat Tham Suea 474
Wat Thammikaram Worawihan 374
Wat Thaworn Wararam 226
Wat Traimit 371
Wat Wang Wiwekaram 227
Wat Yai Chai Mongkhon 206 f., 211
Wat Yannawa 370
Wat Yansangwararam 371
Water buffalo 500 f.

Yaowarat Road 281
Yee Peng Lantern Festival 32 f.

Photo credits

Photo credits for cover images on the flaps:

Alps: Sandra Raccanello/Huber Images; *American National Parks:* Rainer Mirau/Huber Images;

Australia: mauritius images/Masterfile RM/R. Ian Lloyd; *Brazil:* Guido Cozzi/Huber Images;

Canada: Mackie Tom/Huber Images; *Chile:* mauritius images/Michele Falzone/Alamy;

Croatia & Montenegro: Mario Jelavic; *Deserts of the World:* mauritius images/Danita Delimont/Aldo Pavan;

Ice: mauritius images/Masterfile RM/Frank Krahmer; *Iceland:* Getty Images/Arctic-Images;

Ireland: Maurizio Rellini/Huber Images; *Megacities:* Gianni Iorio/Huber Images;

Mexico: Tuul & Bruno Morandi/Huber Images; *Morocco:* Huber Images/Jan Wlodarczyk;

Norway: Udo Bernhart; *Portugal:* Huber Images/Cornelia Dörr;

South Africa, Namibia & Botswana: Getty Images/263Oben; *South Pacific:* Stefano Scatà/Huber Images;

USA: Luigi Vaccarella/Huber Images; *Weather:* Newman Mark/Huber Images

KÖNEMANN

© 2018 koenemann.com GmbH

www.koenemann.com

© Éditions Place des Victoires

6, rue du Mail – 75002 Paris

www.victoires.com

Dépôt légal : 1er trimestre 2019

ISBN: 978-2-8099-1646-1

Series Concept: koenemann.com GmbH

Responsible Editor: Jennifer Wintgens

Picture Editing: Heidi Fröhlich, Katja Sassmannshausen, Jennifer Wintgens

Layout: Marta Wajer/Christoph Eiden

Colour Separation: Prepress, Cologne

Text: Thilo Scheu

Translations: koenemann.com GmbH

Maps: Angelika Solibieda

Printed in China by Shenzhen Hua Xin Colour-printing & Platemaking Co., Ltd

ISBN: 978-3-7419-2028-8